Nature connection is such
physical adventures and m
tionship is vital for our world and it is vital that we care enough
to acknowledge the damage we are doing and take urgent steps to
fix it.

—Alastair Humphreys, author of *Microadventures* and
The Doorstep Mile

This is a book with muscle. Not a softly aspirational book about
belonging and nature but an incisively written work that examines
the needs that humans have for seeing themselves as part of the
natural world. *Reconnection* is an important book that moved me,
made me think and, made me smile.

—Sir Tim Smit, Co-founder and
Vice Chairman of the Eden Project

Fascinating, poignant and hopeful. *Reconnection* should be
mandatory reading for us all.

—Dr Mya-Rose Craig, author of *Birdgirl*

Reconnection is a joy to read! This is a thought-provoking, inspiring
book which highlights the ever increasing need to step outside
and re-embrace the natural world into our lives. For the benefit
of individual wellbeing, for communities and for the health of our
environment, I sincerely hope everybody reads this and seeks a
closer relationship to nature.

—Megan McCubbin, zoologist, conservationist
and TV presenter

As Miles Richardson says, nature makes sense. After reading his
book you too will be in no doubt.

—David Lindo, The Urban Birder

Miles Richardson expertly balances threat with hope in this timely and brilliant book. A must-read for anyone who values the natural world and our connection to it.

—Hilary McGrady, Director General at the National Trust

We're all increasingly aware of how important our relationship with nature is for our own good and for the good of the natural world. Instinctively we know we are not where we need to be. This book sets out in an accessible and thought-provoking way the science that underpins that growing understanding and what we can all do as individuals and as a society to rebuild that relationship before it's too late.

—Beccy Speight, CEO, RSPB

A widening separation of people from nature threatens our physical health, our mental wellbeing and the very survival of our civilisation. In *Reconnection*, Miles Richardson poetically and expertly explores this monumental issue of our time and how we might go about fixing it.

—Ben Goldsmith, philanthropist and environmentalist

Richardson has produced a rich, timely and painstakingly researched account of what's gone wrong in our relationship with nature and most urgently, how it might be fixed. It's never mattered more, has it? I wish every policy-maker, educator, economist and land manager would read and act on this book.

—Amy-Jane Beer, naturalist and author of *The Flow*

Thought provoking, brilliantly researched, and surprising in some of its findings. Also extremely readable which, given the importance of its subject, is helpful for those of us without academic backgrounds. A must-read for educators, policy-makers, and anyone else trying to raise awareness of the benefits and importance of Nature Reconnection.

—Brigit Strawbridge Howard, author of *Dancing with Bees*

Reconnection has the makings of a game-changing classic: hugely sophisticated thought and ideas framed within the most direct and simple language. Any schoolchild could understand it. In fact, all young people and everyone else concerned for the future of life on Earth should read it, if we want to end the nature crisis in our midst.

—Mark Cocker, author of *One Midsummer's Day*

This book is both authoritative and personal, warm and carefully scientific. It busts myths, challenges assumptions and presents truths we can no longer ignore. And crucially, Richardson offers a compelling and practical vision of what we need to do – and why – to change our relationship with Nature. This is the how-to manual and a must-read for anyone searching for the tools to improve human lives and Nature's future.

—Mary-Ann Ochota, broadcaster and anthropologist

It's so valuable to see all the studies brought together and clearly explained – not only as evidence for the instincts we already have about how much connection to nature matters, but also to dispel some myths about how that connection works (or doesn't), and how it might be improved. I found this book absolutely fascinating and I can see it making an important contribution to so many sectors.

—Melissa Harrison, author of *All Among the Barley*

Reconnection is a timely, clear plea to understand just how disconnected we have become from nature. Until it is spelled out, it is easy to assume things are not so bad, that reconnection is just a matter of being more aware. This book shows that the fracture lines go deeper and are more damaging than they might appear on the surface, but it is ultimately a hopeful book, offering solutions that make a greener future seem vibrant and joyful – worth striving for.

—Mary Colwell, author of *Beak, Tooth and Claw*

RECONNECTION

RECONNECTION

Fixing Our Broken Relationship with Nature

MILES RICHARDSON

PELAGIC PUBLISHING

Pelagic Publishing
20–22 Wenlock Road
London, N1 7GU
www.pelagicpublishing.com

Reconnection: Fixing Our Broken Relationship with Nature

First published in 2023
Paperback edition published 2024

A CIP record for this book is available from the British Library

ISBN 978-1-78427-485-6 Pbk
ISBN 978-1-78427-350-7 Hbk
ISBN 978-1-78427-351-4 ePub
ISBN 978-1-78427-483-2 Audio
ISBN 978-1-78427-352-1 PDF

Excerpts from Richardson, M. (2012) *Needwood: A Year in Search of Ordinary Things* and Richardson, M. (2014) *A Blackbird's Year: Mind in Nature* appear with the permission of the copyright holder.

https://doi.org/10.53061/KFSP1540

Contains public sector information licensed under the Open Government Licence v3.0.

Cover design: Jo Walker

Typeset in Minion Pro by S4Carlisle Publishing Services, Chennai, India

Printed in the UK by Short Run Press Ltd

Contents

There is a pleasure in the pathless woods,
There is a rapture on the lonely shore,
There is society, where none intrudes,
By the deep sea, and music in its roar:
I love not man the less, but Nature more.

Lord Byron

Preface

When did my relationship with nature break down? Is my connection with nature now strong? It isn't easy to place our own relationship with nature into the context of others. Those in our family, on our street, in our town, country or continent. In the chapters that follow, we'll find that there do exist ways to measure how connected we are to the rest of nature. For now, it's helpful to consider the relationship with nature that this book is built on. It may chime with many, but describing it will also help those with different backgrounds to understand my perspective. I write as I am.

All my childhood through the 1970s and into the 1980s was spent in the same family home. A semi-detached house where both parents worked. This was of course typical for many; although, thinking back, the large garden and outlook over fields to the front was fortunate. The house was also in a village, with a brook and small woodland at its heart. An ideal location for developing a close relationship with nature, you might think.

I recall moments when lapwings gathered on the playing field at junior school. There was a vegetable patch and a large tree in the playground. I sometimes collected snails on the way home, where they ended up in the garden – a garden with annual flower borders, vegetables and trees. The snails and slugs would be controlled with pellets. Chemicals were used for the 'perfect' lawn. A pump-action pressure sprayer controlled the 'pests'.

Yet starlings still gathered in large numbers, butterflies often fluttered by, blue tits holed milk bottle tops, and occasionally we fed the birds. I enjoyed the birds, annotated a *Collins Bird Guide* and joined the junior branch of the RSPB, the Young Ornithologists' Club.

I recall little nature at secondary school, just a shed full of rabbit hutches. Outside school, hours were spent in the woods, riding bikes, building dens, and some even constructed zip wires and tree houses. Groups of children gathered on the playing field, enjoying football, cricket and thumping each other. Some still went 'nesting' for bird eggs. We crept through gardens, scrumping apples and jumping the hedgerows between front gardens in the 'Grand National'. We were chased, lit fires and enjoyed being outdoors, but there was little to do inside.

Family holidays were short and straightforward. A week in Wales, Devon or the Lake District. The flatlands of Norfolk or the Scottish hills. Time in natural landscapes, enjoying the coast and walking. Although I wasn't a keen walker, and we sometimes came home early as it always seemed to rain. That's my recollection, at least.

Naturally, this childhood was 'normal' for me, and I guess, typical for many. We spent plenty of time outdoors, but did we have a close relationship with nature? Looking back, I don't think so. Nature was the setting. Present but ultimately a sideshow at best, often a resource. Little did we know that nature was in decline. Little have we done to stop that.

The extended time outdoors started to reduce as computers entered every home I knew. An eight-bit world where the eggs were 'chuckie', the camels mutant and hands more likely to be sore from joysticks than nettles.[1] Then I started working. I left school at 16 and went into an engineering apprenticeship – a hollow environment of steel, the scent of machine oil, overalls and intimidation. Nature was absent. I don't recall visiting natural landscapes, and I didn't go on holiday for five or six years. It was a time I waited years to leave.

Joy and restoration came with friends, playing sport and drink-fuelled nights out. Newfangled video jukeboxes, bleached jeans and the latest way to get drunk with sharp ciders or bland lagers. When I was alone, computer games were played, music was bought, and mixtapes were compiled for the car. Nature seemed to have nothing to say.

I quit my job and went to university, aged 21. The first steps towards the present. The built environment still dominated, but there were trips to more natural environments: Bradgate Park, the Devon coast and the Lake District. These were doses of nature I really enjoyed, but looking back, I didn't have an especially close relationship with nature.

I got a graduate job in Manchester, and the after-work culture returned – on a bigger scale and with people who had more to say. There were other differences. Vintage clothes from Afflecks Palace, alcopops, nightclubs and the 3am late bus home. Recovering on Sunday would involve a dose of nature.

This simplistic and concise reflection may suggest I was a clubber, shopper, gamer and socialiser. I was far from any of those. There was no deep relationship with those activities or worlds. Just as there were doses of nature, there were doses of shops, pubs, clubs and computer games. Just as I visited shops and bars, I visited nature too.

At that time, as at most times, this way of living was unquestioned. I was a balancer. Balancing the things one chooses with those one must do to reach some kind of happy whole. Each balancing act can bring wellbeing, but is enjoying nature just another example of this search for equilibrium? Or is it more fundamental to human and nature's wellbeing?

As life moves on, relationships change. Our first house was newly built, and came with a square of turf that we turned into a garden. A typical garden of flowers and shrubs. Chemicals to control weeds in lawns and paths. Little to entice wildlife, other than a tiny pond. Time spent in the garden, walks into the city and around the park, were regular balancing acts.

When children arrive, we can be reminded of the simple wonders we've become accustomed to. We returned to the scene of my childhood. The pram clocked up 1,000 miles in the first year (yes, I fitted it with the mileometer from my bike). Walks to the woods and along the brook. Glimpsing kingfishers and counting ducks (and monitoring their numbers across the seasons). Holidays exploring the coast, lakes, forests and moors. I visited and enjoyed nature a great deal. But I now know that my relationship with nature was limited.

In 2010 that started to change. Over 500 simple, mainly local walks during 2010 and 2011, I experienced a profound reconnection to nature and unwittingly discovered a universal story about our relationship with nature. Somewhat perversely, my first smartphone was the catalyst, the notes app allowing me to write as I wandered, sparking a new engagement with the local countryside. The writing on foot took me on an unexpected journey.

My first steps on this journey took place on 3 January 2010. My notes from that local walk became:

> the sun, our ordinary star, was low in the sky
> and intense in the reflection from the pond that
> had been frozen for a few weeks. It extracted
> and teased the remnants of colour from the
> reeds and I could see little more than the light
> and shade of the mounds of grass on the path
> before me. Attentive, I sensed the faint breath
> of cold air flowing through the hedgerows,
> occasional birdsong could be heard, a lone
> rabbit ran into the field and two blackbirds
> passed through the young trees in front of me.

Unremarkable but meaningful. Reading them takes me back to that very ordinary day. Yet hundreds of those days, thousands of those words, became profound. By 31 December 2010, my experience was very different:

> it was a day suitable to close a year; still, under
> solid skies giving drizzle. I passed a few pigeon
> kills before the likely culprit crossed my path:
> a sparrowhawk, which continued on, suitably
> silent, low through the plantation. Further on,
> rooks were about their nests as jackdaws nearby
> popped and jibed. A magpie departed down
> the ride, an orb of white around its darkness.

I continued along the straight, thin strip, hearing
only the occasional sound of the buzzard. I felt
that I was walking through my mind rather than
the landscape. I slowed to a mindful pace, feeling
each breath, and then a feeling beyond the limits
of my language engulfed me and I wished that
I could plant words as seeds. As the beat of
the rain became apparent, I reached a patch of
old wood pasture where the most ancient oak
seemed to reflect my condition, like a mirror of
the mind. I stopped and I stood in the landscape
until only the landscape remained.

I wrote on foot for another year – a blackbird's year, from its first
song, through its silence and on until its return to song on the
winter solstice of 2011:

I felt an end to my journey; a unity, a
connection sated, an understanding of being
in the landscape. And later, the mild air
brought wonder to the pit of the longest night;
a blackbird returned to song, its energy of tone
defeating the buffeting breeze to enliven every
space and surface. Each phrase a pattern where
thoughts might coalesce and in some mind
make an angel; a blackbird bringing news of a
new year.

Through 90,000 words, I developed a deep reconnection with
nature and increasingly felt part of it. I found myself celebrating
the joy of ordinary things in the natural world. I realised my place
within nature close to home, in the suburbs, villages and fields.
I found that a journey of discovery is not just about travelling to
wild landscapes, but also about finding wilderness in simple places
close to home.

Then, in a new home, the garden chemicals fell out of use,
the pond grew in size, and areas of the garden were left wilder;

untrimmed hedges began to tower over long grasses. For over a decade now my spare time has been spent with nature or managing homes for nature. At work, my focus as a researcher changed; I had walked into the world of nature connectedness. I founded the Nature Connectedness Research Group at the University of Derby to understand people's sense of their relationship with the natural world and create everyday interventions to improve it for human and nature's wellbeing. The interest in the work has soared, resulting in the writing of this book.

The book itself is divided into three parts. After setting out why our current relationship with nature is broken, the first section returns to the beginning, the very beginning, when humans were hunter-gatherers. Once, we were all part of nature – and to understand the present, it helps to look back to when people let go of nature. In trying to do so, we move from the Agricultural Revolution to the Scientific and Industrial Revolutions that formed the foundations for our modern technological lives. Part I ends with some hope, considering the ways we remain connected to nature and introducing the science of nature connectedness alongside my work with colleagues in the research group.

The book's second part explores why reconnection with nature matters – the benefits it brings to both human and nature's wellbeing, and why they occur. And finally, Part 3 considers how we can forge a closer relationship with nature, improving mental health and positive action for our planet. It concludes by considering how the problems set out in the first part might be fixed – and how these methods might be scaled up, so that more and more people can be reconnected with nature.

PART I

THE NEED FOR RECONNECTION WITH NATURE

CHAPTER 1

A Broken Relationship with Nature

A nation of nature lovers?

The British people supposedly love nature. Our greatest poets describe nature's beauty. Our artists depict glorious landscapes. Our naturalists are cherished, and the nation loves nature documentaries that are renowned far beyond our shores. Millions of us are members of nature conservation organisations. Yet research shows the British people have a particularly distant relationship with nature. When 16,000 people across 18 nations were surveyed, it was found that the British people visit nature much less frequently than people in other countries, and their connection with nature was rated much lower too.[1] Despite humans being one of the great apes, only 10% of the population feel that they are part of nature.[2] Nine out of ten people do not feel completely part of the natural world. Perhaps this is no surprise, as it can seem as if there is so little nature here to love and be part of. The UK is one of the most nature-depleted countries in the world,[3] and, as we will see, there is a compelling link to be made between a nation's connection with nature and the abundance of wildlife. The more wildlife a country has, the greater the sense of close connection to nature and the likelihood people will engage with it. Still, even when people developed a closer relationship with nature during the first coronavirus lockdown in Spring 2020, a year later, it had fallen away, decreasing by 25%.[4] There's something about our modern lives that keeps us apart from nature. There is a real need for reconnection.

It's not just the British. The bottom places of the survey of 18 nations, published in the highly respected international journal *Scientific Reports*, were occupied by English-speaking 'Western' nations, all former parts of the British Empire (Canada, USA, Australia, Ireland and Hong Kong). This suggests an attitude towards nature that the British exported and which persists to this day. There is also evidence of an increasing disinterest in nature within modern culture. Our preoccupations and tastes are reflected by the language used in books, films and songs. It has been found that nature words are being used less and less, which reflects certain changes in culture over the decades.[5] While references to nature have declined, we are writing more and more about ourselves, with the use of the word 'me' increasing four-fold since 1990.[6]

Together, through this book, we will consider the broken relationship humans have with the rest of nature – not just in the UK, but in the 'Western' world as a whole, a worldview that is spreading globally.[7] The UN Secretary-General summarised it simply with the words 'Humanity is waging war on nature'; there's now global recognition that solving planetary emergencies requires a new relationship and reconnection with the natural world.[8] We will consider the reasons for the modern worldview that sets us apart from nature, the impact of modern living and why our relationship with nature matters for the wellbeing of both humans and nature itself. Then we will look at how it is possible to form a closer relationship with nature, not just individually, but as a society too.

The state of nature

Before considering the broken relationship between humans and nature and how to fix it, it is necessary to show that it is indeed broken. Some people may take the view that the natural world is not in crisis, that our connection with nature is strong. There are many reports and statistics that show that the climate is warming and that biodiversity is in stark decline. Yet, even here, there is an indication of the lack of interest in the state of nature. There has been a huge disparity in awareness and coverage of biodiversity loss and climate change.[9] Yet only by addressing both the loss of wildlife and the warming climate do we stand a chance of passing on a stable planet for future generations – of people and wildlife.

Climate is the pattern of weather that nature lives within. Few of us, though, love the climate. Instead it is wildlife and its habitats that form the natural world we have a relationship with. So, let's focus on some facts about the loss of wildlife and state of nature globally. These tell us that 69% of animals have been lost since 1970 and that humanity has overseen the loss of 83% of mammals.[10] The landmark health-check of life on Earth by the vital yet rather inelegantly named Intergovernmental Science-Policy Platform on Biodiversity and Ecosystem Services (IPBES) found that the global rate of nature's decline is unprecedented in human history. One million species are threatened with extinction.[11] This continuing loss of wildlife has happened because around two-thirds of the marine environment and three-quarters of the Earth's land has been significantly altered by human actions. This control of land helps create homes and food for humans and our livestock, such that together, by weight, we make up 96% of mammals on Earth.[12] If control and exploitation of natural resources weren't enough, another issue is waste and pollution. Plastic pollution has increased ten-fold since 1980 and that is just one example. As others have shown before, the IPBES report links wildlife loss to human actions.

When one harms another repeatedly for decades, it is at best a failing relationship – but this would better be described as a broken relationship, or even a one-sided war between people and the rest of nature. The global destruction of habitats and wildlife, together with the climate emergency, show that the human–nature relationship is broken. Nature is used and controlled by humans. Of course, this is very general; the relationship isn't broken for everyone. Indigenous peoples have a different relationship with the natural world. However, the IPBES report was a global assessment; the decline of wildlife is widespread, and the human–nature relationship is broken for most people in most places. Since the early 1990s, no country has met the basic needs of its population without over-consuming natural resources.[13] There is a problem even in regions where people are more in tune with nature.

Given this global decline, the UK and English-speaking Western nations must have a particularly broken relationship to present such a low score on measures of nature connectedness. Let's return to Britain. Here, 97% of wildflower meadows have been lost in less than 100 years, and there are 19 million fewer pairs of breeding birds than 50 years ago. Britain has seen nearly half its biodiversity disappear since the Industrial Revolution. The flocks of lapwings that I watched as a child? They have decreased by 80% since the 1960s. The gentle summer purr of the turtle dove is unknown to many with a 94% decline since the 1990s.[14]

These are just a small selection of statistics, percentages, averages from across the globe and closer to home, but let's pause for a moment. Perhaps a minute of silence. Over two-thirds of the animal population has been wiped away in fifty years. That's a catastrophic collapse of fish, amphibians, reptiles, birds and mammals. The wildlife that makes the Earth, and our brief existence on it, so wonderful, so colourful, and so alive – if we pause for a moment to appreciate it. Yet we have let that happen by enjoying the excessive consumption that feeds our modern lifestyles. The urgency of climate change may sometimes be lost through fluctuating weather and the need for projections into the future. With those above figures of 69% and 83%, most wildlife has already been lost. Yet walk through a city and you will see no memorials to that destruction. There is no remembrance of a mass extinction of our own making, an extinction that is set to continue and threaten civilisation itself.[15]

It is difficult to envisage how alive and colourful the landscape once was. I am fortunate to have a wilder and more established garden than many, with plenty of native trees planted by the previous owner in the 1980s. It is a home for nature; brambles spring forth from unruly hawthorns fringed by a border of nettles. It is alive with birds, mice and voles. When I return from the surrounding farmland, it is an oasis of birdsong. Were this narrow patch repeated, between field boundaries far and wide, it would form a vibrant landscape I struggle to imagine. Instead, I find that landscape in the joyful descriptions of everyday nature close to home written by Richard Jefferies, who lived between

1848 and 1887. In his wonderful accounts of time lingering by the fields and hedgerows there's a glimpse of how much wildlife there was on this planet 140 years ago. In *Nature Near London* (1883), Jefferies helps us envision the multitude of birds and wildflowers we have lost:

> A certain road leading outwards from a
> suburb, enters at once among fields. It soon
> passes a thick hedge dividing a meadow from
> a cornfield, in which hedge is a spot where
> some bluebells may be found in spring …
> This meadow in June, for instance, when the
> buttercups are high, is one broad expanse of
> burnished gold. The most careless passer-by
> can hardly fail to cast a glance over acres of
> rich yellow. The furze, again, especially after
> a shower has refreshed its tint, must be seen
> by all. Where broom grows thickly, lifting its
> colour well into view, or where the bird's-foot
> lotus in full summer overruns the thin grass of
> some upland pasture, the eye cannot choose but
> acknowledge it … The thick hedge mentioned
> is a favourite resort of blackbirds, and on a
> warm May morning, after a shower – they are
> extremely fond of a shower – half-a-dozen
> may be heard at once whistling in the elms …
> A pair of turtle-doves built in the same hedge
> one spring, and while resting on the gate by the
> roadside their 'coo-coo' mingled with the song
> of the nightingale and thrush, the blackbird's
> whistle, the chiff-chaff's 'chip-chip', the willow-
> wren's pleading voice, and the rustle of green
> corn as the wind came rushing (as it always does
> to a gateway).[16]

A short walk away from a London suburb to find meadows of wildflowers, nightingales and turtle doves. A little further afield,

Jefferies writes of more colours and numerous insects in *Field and Hedgerow*:

> This forest land is marked by the myriads
> of insects that roam about it in the days of
> sunshine. Of all the million million heathbells –
> multiply them again by a million million
> more – that purple the acres of rolling hills,
> mile upon mile, there is not one that is not
> daily visited by these flying creatures. Countless
> and incalculable hosts of the yellow-barred
> hover-flies come to them; the heath and
> common, the moor and forest, the hedgerow
> and copse, are full of insects. They rise under
> foot, they rise from the spray brushed by your
> arm as you pass, they settle down in front of
> you – a rain of insects, a coloured shower.
> Legion is a little word for the butterflies; the
> dry pastures among the woods are brown with
> meadow-brown; blues and coppers float in
> endless succession.[17]

If only we could journey back in time to see how our increasingly sterile landscape was once alive. In places, we can. There are many nature reserves and one of the few positives in our modern environment is the growing number of rewilding projects, with around a quarter of a million acres rewilded in the UK. However, this is a tiny 0.4% of the country.[18] The stark reality is that the UK has one of the worst records for biodiversity in the world. While only pawns within a wider system of destruction, most individuals do very little to improve habitats and biodiversity. Indeed, many actively harm nature, the scale dependent on their wealth, from grouse moors to gardens. Garden centres actively promote dozens of insecticides and weedkillers, marketed as 'garden essentials' under the banner of 'love the garden' to help 'protect and enjoy garden moments'.[19] Rather than being in decline, the $7 billion global market for home and garden pesticides is expected to grow

5% a year through to 2028.[20] This trend of domestic control and removal of nature is reflected in the sharp rise in artificial lawns. Online searches for 'artificial lawns' increased 185% in May 2020, and the $2.5 billion global market is predicted to grow to $5.8 billion in 2023.[21] These products are also a step towards 'ecological simulation' where natural ecosystems are replaced with simulacra, stylised representations of nature that satisfy cultural desires but remove actual nature and interactions with it.[22] This cultural desire for green spaces is often as a resource for human activity, which itself can cause further damage. For example, although attracted to green spaces and beauty spots on sunny days, there is also a growing problem with littering. A total 300 tonnes of rubbish were removed from Brighton beach alone during the summer of 2021, with 30 tonnes of litter accumulating over one weekend.[23]

Why is there a seemingly deep desire for nature yet little care for it? Over the next few chapters, let's take a look at our societal and individual relationship with nature and ask some challenging questions along the way. About our gardens, our pets, our approach to wellbeing, even the stories we tell our children. About the technological Western worldview that forms the way we see the natural world. To do so, we also need to look further back, to ask the question: have people ever lived in harmony with nature? And if we once did, when did our relationship with nature start to fail?

Understanding our broken relationship with nature

A simple observation provides a starting point. Some relationships with nature are positive. Some are negative – those relationships that have led to the current state of the natural world. Back in 1993, Stephen Kellert, a professor of social ecology at Yale, produced a list of types of human–nature relationships in his work on the values of biophilia. Biophilia refers to an innate affinity for living systems, and the connections humans seek with nature for both survival and personal fulfilment.[24] It is important to note that although biophilia is often thought of as meaning a love of nature, it includes relationships with nature that can be positive and negative. Clearly humans, like any other animal, need to use natural resources for survival, but as we've seen, that can go way too far.

So, back to Stephen Kellert. His list of nine values or dimensions of biophilia is easier to think of as a list of types of relationship with nature. In research we'll consider in more detail later, we found that four of these types of relationship were unrelated to a closer connection with nature. These four were dominion over nature, utilitarian use of nature, fear of nature, and a scientific relationship with nature. These types of relationships are common, with utility and, at times, fear essential for human survival. However, as we've seen, use and control of nature have accelerated. Indeed, they have been industrialised.

In capitalist societies dominated by Western thinking, this use and control of nature is often undertaken with a promise of prosperity and happiness. Our use of and increasing dominion over the natural world, rather than existence within it, has been seen as a sign of human progress. The success of the Scientific Revolution, Western civilisation and its culture being seen as the pinnacle of evolutionary progress.[25]

Yet that progress seems to be linked to a more distant relationship with nature. If we return to the survey of 18 nations mentioned earlier, the nations of Western Europe and the Western offshoots of the USA, Canada and Australia – those countries with more distant relationships with nature – have seen particularly fast levels of growth since 1820, with personal income increasing thirteen-fold and seventeen-fold respectively. This is compared to ten-fold in Southern Europe and six-fold in Eastern Europe. On other continents, growth was three-fold in Africa, six-fold in Asia and Oceania, and seven-fold in Latin America.[26] It seems that as we consume to feed our modern lifestyle, we become more disconnected from nature. Affluence increases resource use and pollution, and our societies, competitive market economies and cultures are fundamentally based on growth that incites further consumption.[27] Put simply, growth means more: more wealth, more consumption, which – especially in the absence of progress based on quality – comes to define our own growth as human beings.[28] We become consumers.

The utilitarian use of nature is such that every country that meets the basic needs of its population is overconsuming natural resources.[29] For example, humanity consumes an Iceland-size area of forest each year.[30] Similarly, our dominion over nature is such that globally we spray over two million tonnes of pesticides annually and regulate the flow of two-thirds of the longest rivers.[31] The use and control of nature destroys habitats and is driving species to extinction. Indeed, this utilitarian and dominionistic relationship with nature has affected the planet to such a long-lasting and permanent extent that humans have created a geological epoch, the Anthropocene. When use and control of nature dominate, a broken relationship with nature follows. Further, as we'll see later, an overly scientific relationship with nature can be problematic too. But first, let's travel way back and consider why this control and use of nature dominates the modern human–nature relationship – why the Inuit refer to Westerners as 'the people who change nature'.[32]

The letting go: from hunter-gatherers to farmers

For over 90% of our history – that's around 10,000 generations – humans existed as hunter-gatherers.[33] Each generation successfully reproducing to allow you to read this book today. Despite what today's glossy adverts and influencers might say, their fundamental needs were no different to those our own. They ate, needed shelter, and had families in egalitarian, often nomadic, social groups. Survival was dependent on understanding animals, both for hunting and escape, and this will have informed their worldview. Just like us, hunter-gatherers had a worldview, a fundamental framework of ideas and beliefs. This was different to ours, with studies suggesting that everything was connected in the worldview of the hunter-gatherer, from that which is clearly visible, such as rivers, trees, soil, insects, hills, animals and people, through to ancestors and spirits in an integrated and dynamic whole.[34] Hunter-gatherers were likely animists. Rather than viewing themselves as superior to other creatures, they would have viewed humans and animals as one, and all life as having a role to play.[35] So, for most of human existence, we did not see ourselves as different to other animals; indeed, for hunter-gatherers, animals may have been equals, superiors or worshipped. Although most people are no longer sure they are part of nature, our long history shows that 'a feeling of belonging with the rest of nature is normal'.[36]

Animism and the belief that objects, places and natural things as varied as animals, plants, rivers and stones have a spiritual essence or soul is a relational worldview of many Indigenous peoples.[37] It is still seen in surviving hunter-gatherer communities – for instance, the Nayaka people of South India see animals and plants as sentient, and they talk to the animals.[38] For the Co Tu of Vietnam, trees have great importance; they have souls and are personified as gods, creating a culture of protecting forests and trees.[39] From a modern perspective, animism is often grossly misunderstood as a simple religion and failed worldview.[40] Yet today, we're failing nature and at the same time 'relationally framing' the technology we use: naming our cars, saying 'please' and 'thank you' to our digital home assistants, personifying the things we're close to – the objects of the concrete jungle, rather than the forest.

Animism is based on a 'relational epistemology', or to put it more simply, thinking in terms of relationships. This is built upon the form of knowledge that shapes how people know an 'other' in order to live *with* that other, rather than simply knowing facts and figures *of* that other.[41] In animism, beliefs and actions nurture and focus on living mutually with many others, be they people, animals or things. This relational epistemology contrasts with the dominant modern theory of being where these others are separate objects we may study or control.

It is thought that such relational worldviews often included the notion of 'nature as parent', where the forest that provided food and shelter is mother or father.[42] Just as we trust our parents, hunter-gatherers trusted the forest and wider natural world to meet their needs. Just as parents give, and at times take away, so does nature. However, Indigenous peoples, including in more recent times, understood that the giving natural world must be repaid through sustaining the land. For example, the Native American Blackfoot tribe have a buffalo legend which illustrates the equality between humans and wildlife through a story of cooperation and reciprocal life.[43]

However, this harmonious relationship with nature didn't exclude hunter-gatherers from the shifting baseline syndrome, where knowledge of the former state of nature is lost and each generation accepts the current state as normal. Hunter-gatherers were certainly less aware than modern-day humans of any long-term negative impact of their activities. Although debated, there is a body of evidence that suggests their use of natural resources through hunting led to extinctions of megafauna.[44] Small populations of hunter-gatherers such as Polynesians in New Zealand wiped out megafauna such as the moa in short timescales through hunting and loss of habitat.[45] The exploitation rate we exert on other species is thought to be up to 15 times higher than for any other predator.[46] Can humans ever truly live in harmony with the rest of nature? Although some Indigenous peoples have done so for long periods, that may be largely due to low population. When our population and technology 'succeeds', it tends to bring ruin for other animals.[47] Perhaps harmony can only come when people have and act on feedback on the negative consequences of their actions, however unpalatable that may be. We now have the feedback, but the signs of action are far from encouraging.

Clearly, hunter-gatherers were not mass consumers; the populations were very small, and nature provided what was necessary, although they managed the land to increase natural bounty. However, when Westerners discovered indigenous societies that remained after the Agricultural Revolution, they were deemed to have no right of possession as they were not 'working the land' – rights were based on use and control.[48] Some expeditions found that hunter-gatherer societies did not value the Western artefacts they were given and, somewhat controversially at the time, the anthropologist Marshall Sahlins suggested that hunter-gatherers had an 'affluent society' with unburdened and abundant leisure time.[49] This differs markedly from the modern conception of affluence built on a consumer economy that meets the 'needs' of people. Many of those needs are created by the market and technology. As we'll see, recent thinking into a close relationship with nature as a basic psychological need taps into a core aspect those early societies had, but that we have lost.

Studies suggest that, although valuing hard work, hunter-gatherers spent only a few hours a day on basic needs such as collecting food, and days could be enjoyable and unhurried and always spent embedded in the natural environment.[50] Whereas now 93% of our time is spent indoors or in a vehicle, with urban dwellers in the UK typically spending less than five minutes in green spaces each day.[51] Given the nature of urban green spaces compared to wild forests, the notion of embeddedness in nature is very distant. Embeddedness is now an aim of technology instead, with immersive augmented reality ostensibly offering wellbeing as well as profit.[52]

Another difference between hunter-gatherer and modern societies comes in the form of food. We no longer need to forage; our food is purchased and stored for a few days or weeks into the future. Although preserving some meat, hunter-gatherers lived much more in the present, trusted nature to deliver and shared their food within their social group.[53] Since the Agricultural Revolution, people have worked hard to invest in the future. There's less willingness to share the products of that hard work. When there is even a suggestion of uncertainty about supply, for example before coronavirus lockdowns, people panic-buy to build up their own stores. Rather than nature, our trust is now in supermarkets and their supply chains.

The shift to farming and agriculture was patchy and slow, yet a massive revolution in the lives of humans. Jared Diamond describes how the Agricultural Revolution was a catalyst for many changes, from ownership and possession and wars of conquest, to writing, maths and accumulation of wealth. It saw slavery, epidemics and famine, such that early farmers suffered more than their hunter-gatherer ancestors.[54] Whereas hunter-gatherers had a varied diet, crops were less varied, and human health deteriorated as a result. Archaeological evidence suggests that in fourth-century Britain, after many centuries of farming, half the people were dead by the age of five and most didn't reach 40 years old.[55]

The long and drawn-out Agricultural Revolution is when humans started to let go of their close bonds with nature. Agriculture saw the end of nature as giving parent. Where food had once been a gift from the Earth, farmed food now resulted from hard work.[56] This way of living led to new values and a culture based on accumulation and dominion over wild places as the landscape became divided and owned. As we'll see in more detail later, even now the differing way nations farm their land impacts on individual relationships with nature. The Agricultural Revolution also opened a new chapter in human relationships with animals. This included the domestication of 20 species that suffered the costs of this new relationship, while the remaining wild animals started to have their habitats removed.[57] The anthropologist Tim Ingold explains that dominionistic relationships emerged with the adoption of agriculture, in particular cultivation and animal husbandry.[58] At this point, animals, plants and even land became objects – the 'who' became 'what'.

Through studying the Nayaka, anthropologists Danny Naveh and Nurit Bird-David explain how the relational worldview of animism, still seen in indigenous cultures of modern times, was being lost as humans became involved in farming – the animals were becoming objects.[59] However, to the Nayaka, the animals of the forest were still persons and co-dwellers, sentient beings able to feel and think. This is a belief system seen across hunter-gatherer cultures, where a relational epistemology is common.

Naveh and Bird-David explain that this transition from co-persons or dwellers to things occurs when animals and plants become objects with a purpose; it is not caused by dominion alone. Their research found that the socio-economic context influenced the human–nature relationship. The more remote the feedback to the consumer – the greater the gap between action and consequence – the less care and empathy there is. When a way of living places a distance between the consumer and consumed, it fractures the relational outlook where wildlife and people are one.

Interestingly, research has shown that children demonstrate animistic thinking, considering some things to be alive, especially if they move, and this decreases with age – although it does not disappear entirely.[60] This can lead to children and some adults thinking that objects, recently robots, have life, soul or agency. However, in this research, animism is typically framed as an 'error' rather than freedom from rationalistic schooling, something to be nurtured and listened to if we are to form a closer relationship with nature.[61]

This new dominionistic relationship with animals that grew from the move to an agricultural way of life was justified by theist religions, which focused on humans and gods. Many of the great religions emerged when humans moved from hunter-gatherers to an agricultural civilisation – and as the plants and animals fell mute, conversations were had with gods instead.[62] These conversations were needed for other reasons too. For hunter-gatherers, nature was the provider. With the move to agriculture, nature became the force that could take away through the weather. Of course, the weather can't be controlled by people, but can be controlled by the gods.[63] The new-found control over the landscape was accompanied in most agricultural societies with the belief that gods manage the weather, still reflected in the insurance term 'act of God'. Appeasing the gods could indirectly influence the climate.

Rather than the hunter-gatherers' woodland spirits within a dynamic web of life, gods became dominant divine forces overseeing, and hopefully approving of, human activities. Nature became an opposing force appeased via gods. Further, with the advent of static communities, land became valuable, with other tribes seeking ownership of fertile areas. This provided the context for dualistic religion based on good and evil, a concept that placed further distance between people and nature.[64] For instance, Zoroaster was an ancient Iranian prophet who taught a dualistic cosmology of good and evil, and some wildlife was seen as noxious or evil.[65] Such thinking is thought to have informed later religions that gave humans dominion over animals.[66]

Although farming changed the human–animal relationship and a form of dualism developed in religions and beliefs, material and spiritual phenomena were not clearly divided. Up to the sixteenth century, people in Europe still had a sense of belonging to their natural environment. Their worldview and the fundamental way people perceived themselves and their relationship with nature were radically different to ours today.[67] People still saw trees, rivers and the landscape as enchanted, alive and wonderous.[68] Yet, as we shall see, this organic worldview and relationship with nature was split surprisingly quickly by the shift to modern dualistic thought, where the rational mind became separate from the mechanistic body. Animals were viewed as machines and humans set themselves apart from nature. People became more interested in themselves than in the wider world.

CHAPTER 2

The Great Theft

A revolution in the way we think

While humans started to let go of their close relationship with nature as early as the Agricultural Revolution, the more recent shift to modern ways of thinking can be seen as a great theft. As humans loosened their bonds, nature's place in our lives was eventually snatched away. The shift in thinking in Western Europe during the seventeenth century, when the mental and physical were split into two separate concepts, was profound. Modern self-awareness was born, alongside individual self-interest and a way of thinking that made animism seem irrational. Nature became separated; the human relationship with it fractured. Yet this move to dualistic thinking was also improbable, being described as 'a near-miraculous concatenation of circumstances'.[1] Of course, there is a long history of dualism, such as the separation of mind and body in early Greek philosophy, the traces of which can still be seen in Western thinking today. Plato saw the mind and body as made of different substances, with the rational mind being the source of legitimate knowledge and truth, whereas the body and senses were not to be trusted.[2] As we'll come to see later in the book, losing trust in the body and the senses is a poor foundation for a close relationship with nature.

As it's such a profound part of the story of the human relationship with nature, let's look at dualism a little more closely. The key transition in dualistic thinking begins in the seventeenth century with René Descartes, the French philosopher, mathematician and scientist described as 'the father of modern philosophy'.[3] His mind–body theory included the famous statement translated to, 'I think, therefore I am'. The irony is that the dualism that says we exist inside our heads was perhaps based on a hallucination Descartes had when stuck in a steam room![4] The theory, and statement, suggests that reason and the mind reign supreme as the core essence of human existence, rigidly distinct from the body, which is seen as unthinking matter. This change, where the embedded worldview was replaced with reason and individualism, was dramatic.[5] It was a shift in human consciousness – from a perspective where the person and what they observe are one, and the rest of the world can be known through that relationship, to one where the world 'outside our heads' is known through knowledge and thinking 'inside our heads', through information delivered via separate bodily organs used to perceive the world. This way of thinking swept away the relational worldview that embedded people within nature.

As reason gradually saw the retreat of religious dogma, this influential view that mind and body are separate and the mind is what makes us human underpinned the rise of the modern scientific worldview and the Age of Enlightenment in the seventeenth and eighteenth centuries.[6] This 'enlightenment' provided the basis for the Scientific and Industrial Revolutions, with their material progress spreading and reinforcing the dualistic worldview that has since had a massive impact on the natural world. It wasn't just primacy of mind and reason; Enlightenment thinkers aimed to replace religious faith with faith in humanity.[7] Humans became the overlords of nature. The psychological impact of this transformation is arguably at least as important as the Scientific and Industrial Revolutions themselves. Although fundamentally changing the way people saw themselves and their place within the natural world, the transformation can be seen to have liberated people from magical and irrational beliefs. Although, ancient people were not simple-minded. Science and mathematics were in existence and part of Indigenous knowledge centuries earlier and were likely incorporated into 'modern' science.[8] Ancestral ways of knowing can be a source of great wisdom with Indigenous peoples' critique of power, money and equality in European society creating a backlash amongst enlightenment thinkers.[9] Still, the

Enlightenment was a fundamental event providing a new clarity of understanding that released technological potential and a material world of production and commerce that has transformed our world.[10]

It is no surprise that this profound shift in thinking is still fundamental to the modern human–nature relationship. When mind is identity, bodies have no intrinsic value, whether they be human or animal. Descartes argued that animals were mechanistic and denied they could feel pain, views that, although not universally accepted, provided justification for humans to mistreat animals. Further, without mind, nature has no intrinsic value other than to be studied by the human mind – studies that centuries later would show the sentience of many animals and that even plants can learn.[11] Hopefully, attitudes are now changing, but when nature became an assortment of mechanical objects existing outside us and valued by their potential for exploitation, the essential human–nature relationship was broken.[12] No longer were humans and nature one and the same.

Another influential figure of the Scientific Revolution was Francis Bacon, who is credited with creating the scientific method. Descartes wrote of science allowing humans to be masters and possessors of nature. Francis Bacon set a similar tone for scientific investigation, asserting that humans have power, command, dominion and rights over nature by divine request.[13] The foundations of modern dualistic science were formed and soon advanced by influential scientists such as Isaac Newton.

The Industrial Revolution brought Bacon's vision to fruition and changed the human–nature relationship.[14] Human ingenuity exploited natural resources, such as coal, to fuel the Industrial Revolution and 'the human urge to break free of the bonds of nature'.[15] Each new technology gave greater control over nature and allowed greater exploitation.[16] The increasing efficiency rewarded and funded further development, as there was money to be made. The steam engine's power only started the Industrial Revolution because it made a profit.[17] As factories allowed massive increases in productivity, European empires extended as there was a whole world of natural resources and Indigenous peoples to exploit to accumulate wealth. Commerce would conquer the constraints of nature.[18] Through the Agricultural, Scientific and Industrial Revolutions the control and use of nature accelerated and a perceived right to exploit nature took hold.[19]

Of course, this change in thinking and employment was accompanied by a change in location. The Industrial Revolution led to urbanisation. People left their villages and rural agricultural lives to work in factories – an extinction of experience that intensified the split from nature. The contrasts between rural and urban living helped feed the Romantic Movement and its appreciation of nature, with dark and foreboding wild places increasingly revered by tourists and the harms of urban living raised.[20] Indeed, Romanticism was a reaction against the Enlightenment and the Industrial Revolution, with William Wordsworth being a founding figure. Wordsworth's poetry captures many of the themes in this book. He rejected the human dominance of nature and saw how the relationship between humans and nature was ruined by the Industrial Revolution, destroying the natural world.[21] Wordsworth saw how the emerging modern age disconnected people from nature, people who then became egocentric.[22] We celebrate the poets that resisted this vision of industrialisation, but their defeat is a story not often told.

The scientific and rational dualistic mindset led to remarkable scientific and technological progress, transforming the lives and health of hundreds of millions of people for the better. Yet that mindset placed humans apart from nature, justifying its control and exploitation. It was felt that science would command the world and bring happiness by meeting all human needs.[23] Material progress has been closely linked to happiness, and these two notions dominate the Western worldview, which has increasing global influence.[24] For example, it is suggested that the African environmental crisis grew from adopting the Western worldview which was spread through science, technology and industrialisation.[25] When considering material progress, the world has done exceptionally well over the past 200 years, with substantial improvements in living standards and life expectancy. However, ultimately there's no human wellbeing without nature's wellbeing. Without a healthy planet, those gains will come tumbling down.

The power of language

Humans being the masters and possessors of nature was a popular viewpoint and metaphor.[26] For instance, Sigmund Freud wrote of humans coming together and 'taking up the attack on nature, thus forcing it to obey human will, under the guidance of science'.[27] Such metaphors and their accompanying scientific conquests justified and enabled ever-increasing control and exploitation of nature, and humans inflated themselves to gods.[28] These metaphors are powerful; they seep into everyday discourse, frame our thinking, forming concepts of truth and reality.[29] Control and use of nature became a mainstream way of thinking, absorbed into society and creating illusions of separation and supremacy over nature in everyday discourse, in our leisure, work and disciplines of knowledge. As just one example, the science of psychology has been seen as 'mute' on the environmental crisis and as normalising damaging behaviours and lifestyles.[30] More recently, as a focus on human behaviour and its impacts on nature became a more common though still niche focus, it was noted that, unsurprisingly, psychologists are more interested in people than the state of those people's home, the natural world. There was a need for psychologists to apply the extensive methods and toolkits of psychological constructs more effectively to nature conservation issues.[31] As we'll see later, there is still a surprising disinterest in studying the human–nature relationship, yet it is fundamental to our future.

The powerful metaphors that spoke of human control over nature set the context of our worldview. Metaphors produce mental structures or frames that shape how we see the world. They have a profound influence on how we think and act. Even a single word can initiate a metaphor that can change how people think and gather information, yet many do not recognise the influence of metaphors in their decisions.[32] George Lakoff tells us that strong metaphors even define what we see as common sense.[33] Lakoff says that our perceived separation from nature is so deep within our conceptual system that it is difficult to overcome. The separation of humans and nature in everyday language and discourse creates real boundaries to a close relationship with nature. In this context, finding cultural harmony with nature is exceedingly difficult. If everyday speech and metaphors suggest that nature is an *other*, separate from humans, a resource to be controlled and exploited, this will inform our viewpoint and actions. Deeply ingrained views are at the root of climate change and wildlife loss; otherwise, such global-scale disasters that threaten civilisation would not have been allowed to happen. Nature is no longer the giving parent; nature has been more akin to an expendable slave, there only to serve its human master.

The dualistic metaphors and scientific pattern of Western thinking promoted abstract reasoning about the natural world. Nature could be explained by fixed laws and neat mathematics. Little wonder that Kellert's scientific dimension of biophilia was unrelated to a close relationship with nature. Each day we are bombarded with science, discovery and progress. Each day a little more of the unknown becomes known. Modern science divides humans, as observers, from the observed natural world. Nature becomes the object of analysis, described as abstract matter obeying laws. Essential knowledge, once passed from parent to child in the landscape through talk, is replaced by written facts delivered in the classroom and in scientific communities. Previously, parents' oral traditions necessarily put nature in the foreground through close observation of nature, teaching children to notice interrelationships. Even in recent decades, when speaking of outdoor practices, parents in indigenous communities talked of foraging, forest walks, fishing and medicinal plants. Whereas for European-American families, outdoor practices were sports, and activities such as cycling and canoeing that moved through nature as a background.[34]

Objective analysis of the outer world and the science of the rational mind are seen as the source of truth. Our inner subjective experiences are often dismissed as irrational and inferior. Yet we strive to satisfy them with 'truths' from the outer world, consumer goods that have become the indicators of success.[35] Modern work is dominated by creating these consumer goods, and forms of recreation where experiences are created, packaged and served with, at best, nature as a background.

Turning inwards: the birth of humanism

The Scientific Revolution was a key chapter in human–animal relations. A divide was placed between them, the bond to nature broken. This fracture also raised the status of humanity. The gods began to be silenced through human understanding of the laws of nature, and humans became gods instead with their accompanying angels of new technology. Rather than appeasing the gods, human needs and desires now mattered most. Humanism was born – a human-centred approach to life fuelled by science and dualistic thinking.[36]

Humanists 'believe that human experience is the supreme source of authority and meaning'.[37] They trust in reason, autonomy and science for the purpose of discovering truth and are committed to the interests of humans.[38] Humanism suggests that human flourishing is maximised through happiness, health, knowledge and rich experiences.[39] It includes the sanctification of human emotions, desires and lives, leading to a quest for power and immortality. Humanism comes from within us – our emotions and feelings, rather than spirits in the forest or angels and demons. Humans decide what matters and have faith in the human species. It is humanism itself that brings meaning to our lives and the cosmos. Add in the separation of humans and nature, and we can easily see how the meaning and importance of the natural world became so diminished.

Humanists celebrate the expression of the unique individuality of each human, their feelings and traits and quirks, be that of themselves in liberal humanism or of others in socialist humanism.[40] Humanism is a worldview where labels are resisted, and multiple identities acknowledged.[41] While individualism may be encouraged, however, one is still conforming to humanism.

For humans, our belief in self and individuality defines us as separate from each other and other life forms. Under humanism, our values are based on human desires; the rest of nature can have no value without humans. Confirmed by the daily experience of the latest technology, humanism is a science-fuelled belief in progress beyond the limits that restrict other animals. It is a near-universal but groundless faith in the idea that the human species can control its destiny and be free.[42] The accompanying surety in and focus on our individual selves keeps us apart from nature.

That covers the birth and core aspects of humanism, but you may be wondering about recent evidence for this belief in self and the spread of individuality. The rise of individualism has been studied by Jean Twenge and colleagues. They found a 42% increase in personal pronoun use between 1960 and 2008, with individualistic phrases also increasing.[43] Similarly, in the 1980s, songs switched to using 'I' more than 'you'.[44] Where people used to be more inclined to think and sing about others, they started to sing about themselves. Nowadays, Western culture favours an individualistic identity and there is an emphasis on the expression of individualism and independence in the pursuit of self-fulfilment. This is accompanied by a universal assumption that Western values apply across the whole world.[45] Modern life and mass consumption is creating an epidemic of narcissism that is making us more self-centred with an excessive interest in our own needs.[46] Researchers are now exploring 'collective narcissism' where whole in-groups of people believe they are not receiving sufficient recognition and deserve special treatment and validation. This is associated with superficial virtue and hostility towards those outside the group.[47] Perhaps 'I think, therefore I am' is truer now than it has ever been, and that could be bad news for nature. The modern conception of self, an epidemic of narcissism and people viewing themselves as superior and separate from the natural world are major barriers to solving environmental problems.[48]

With such a powerful belief in the self and individuality throughout Western culture, it's no wonder it feels so real and can become addictive. William Van Gordon and colleagues provide an enlightening account of 'ontological addiction theory' and 'emptiness'. These ideas stand at odds with the dominant Western worldview, so can be tricky to grasp and explain. I've tried to keep it simple, and I hope by the end of this paragraph you will see the value of it. Ontological addiction theory suggests that 'human beings are prone to forming implausible beliefs concerning the way they think they exist, and that these beliefs can become addictive leading to functional impairments and mental illness'.[49]

Ontological addiction is the unwillingness to abandon the belief in the 'self' or 'I'. It is a Buddhist perspective where things do not exist independently, which includes the self, the 'me' inside our heads.[50] Rather than existing independently, everything, including humans, exists in a relative sense and is dependent on all other phenomena. The Buddhist term for this perspective where things are 'empty' and dependent on interconnectedness with others, is 'emptiness'.[51] Within this philosophy, a tree exists in reliance upon an ecology of many things such as water, which in turn is reliant on rain, such that if one component were absent, the tree would not exist. Likewise, the tree is essential to the existence of other phenomena. The tree is inherently 'empty' of a self but 'full' of the natural world.

The view that the independent self does not exist challenges Western psychological theories that suggest the self exists and is important for our self-image, self-worth and wellbeing.[52] At the centre of our existence, this powerful sense of self creates a world where the rest of life is peripheral.[53] When nature is peripheral, interactions with it can reaffirm selfhood and further secure the sense of self, creating a feedback loop. This leads to the existence of the self becoming a firm belief, silencing the idea that we are one integral part of the wider natural world. With a firm belief in the self, the aim of an activity is likely to be to satisfy oneself, when in actual fact we are 'empty' without the rest of nature. The self consequently becomes a 'thirst that can never be quenched', and there can be no lasting happiness, just pursuit of things. This focus on 'me, mine and I' eventually restricts the awareness of the present moment, which, as we'll see later, can help build a close relationship with nature. Understanding and cultivating 'emptiness' helps form a more accurate view of reality where the self is inseparable from the natural environment – when all things are empty, the world becomes full.[54]

To help explain a little more, Charles Eisenstein provides a useful account in his book *The Ascent of Humanity*.[55] Propelled by science, the spread of the 'I think, therefore I am' philosophy of René Descartes created the independent 'Cartesian self': the 'me' that is fundamental to modern life. Yet evidence from Buddhism, but also neuroscience, suggests it is an illusion and does not exist inside us. Just as we are biological creatures within a wider natural world, our feelings of mind and meaning come via our actions in the wider world. So, me-ness, if there is such a word, emerges from relationships with others; the key point being that none of us are separate individuals, and we all depend upon each other – without you, there is no me. Eisenstein suggests that rather than life being about 'survival of the fittest', where one species or individual is set against another, it is more cooperative. What matters more, as we will see in Chapter 4, is relationships. The relationships within nature are a wonderful thing. It is interesting to note that in *The Descent of Man*, Darwin regularly used the word 'love' and explained that he had previously attributed too much to the term 'survival of the fittest' and he was not referring to the victory of one over another.[56] Cooperative and close relationships fit well with the earlier accounts of nature as provider. The Industrial Revolution introduced a race to consume natural resources – such that modern life can feel like a competition to know more, to walk further, run faster, climb, cross, conquer and consume. Hence, the misuse of the term 'survival of the fittest'. The different perspectives on the self, introduced above, can highlight alternatives to the dualistic model of a mechanistic world.

Alternative philosophies

While Western thinking was driven by a dualistic outlook and the search for laws that govern nature, Eastern philosophies focused on harmonious principles based on the natural world being one entity. A world in which humans were more embedded rather than separated – thinking that is now consistent with recent science that is finally taking steps away from mechanistic processes and towards integrated biological systems.[57]

Chinese thinkers rarely mistook ideas for facts, and they saw how nature worked harmoniously and effortlessly, the days and seasons came and went, and the river flowed.[58] Philosophies such as Taoism extolled working with the flow of nature rather than purposeful action against it. In *The Patterning Instinct*, Jeremy Lent offers a rare insight into the philosophy of the Chinese Song dynasty, which, he argues, is highly relevant to modern environmental challenges. Before and after the Norman Conquest of England in 1066, Song thinkers linked mind and body, reason and emotion. This coherent theory used terminology that is lacking in English. The heart and mind, *Xin*, were combined as the site of feeling and reason, with that relationship extended to all living systems within a universe governed by principles and dynamic patterns of cohesion, termed *Li*. An understanding of *Li* was a sense of love, *Ren*, arising from the engaged realisation of connectedness to all other beings. Human's spiritual goal and destiny was not to transcend or repress emotions but to honour and harmonise them with the wider natural world, and realise and feel part of it. *Ren* was a route to optimising one's own life and existing in harmony with others and wider nature.[59] As we'll see, this destiny parallels the modern psychological construct of nature connectedness. Such harmony of emotions can also be compared to current models of affect regulation that explain how nature helps to manage our moods and keep us well.

Although this Neo-Confucian worldview may seem unscientific, Lent explains how this dynamic interconnectedness and being within a larger whole is seen in modern systems biology, from cells to ecosystems. Indeed, although dualistic thinking dominates our everyday outlook and discourse, well-known thinkers concur with the more holistic approach. Albert Einstein wrote:

> A human being, is part of a whole, called by
> us 'the Universe,' a part limited in time and
> space. He experiences himself, his thoughts
> and feelings as something separated from the
> rest – a kind of optical delusion of his con-
> sciousness. This delusion is a kind of prison
> for us, restricting us to our personal desires
> and to affection for a few persons nearest to us.
> Our task must be to free ourselves from this
> prison by widening our circle of compassion
> to embrace all living creatures and the whole
> nature in its beauty.[60]

The Song dynasty fell to a Mongol incursion in 1279. The Neo-Confucian worldview did eventually become established in the Chinese state by the sixteenth century. However, over time some of the pioneering thinking became lost, and the philosophy was blamed for the Western colonialism of imperial China.[61]

More harmonious worldviews can be found elsewhere in the world. The Bishnois, a Hindu community in India, have been referred to as 'born nature lovers', and South Asian and Indian religions also suggest greater unity between human and nature's wellbeing through the concept of Dharma.[62] Even when faith has dualistic elements, there can be a strong ethos of nature conservation, with nature being held in high regard. For example, in the Bahá'í tradition, there is a distinction between matter and spirit, yet also a focus on unity and oneness of all things.[63] Rather than dualism being fundamentally flawed, it is perhaps the dualism of the dominant Western worldview that is problematic.

There were alternative philosophies in Europe too. Spinoza offered a more emotional and relational understanding that combined God and nature, mind and body.[64] In several respects, the German philosopher Arthur Schopenhauer was unusual, but notably for us in that he absorbed Indian thinking and its core principle that the free individual is a mirage that conceals reality. Schopenhauer saw that we are at one with other animals, driven by bodily needs in a world where everything is related.[65] The Romantic movement also rejected mechanistic metaphors and saw a world alive with emotions rather than functional objects. However, the strength of that argument perhaps furthered the divide between emotion and reason. Nature lovers can be seen as romantics, forever in confrontation with rational mechanists who seek to remove emotion from learning about nature.[66]

For some, nature, as a dynamic living system with many interconnections, could never be fully understood. Such ideas can be seen in the formation of phenomenology in the early twentieth century. Phenomenology rejects scientific objectivity and recognises humans as embedded in the wider natural world. Merleau-Ponty was a French phenomenological philosopher who highlighted the interconnection between the perceived and perceiver, and how humans are embedded in the landscape. He wrote of the human body as 'the vehicle of being in the world' and 'the Flesh' as a collective term for the flesh of the human body and the flesh of the wider world.[67] Phenomenology moves us away from a goal-directed focus to a holistic worldview without our boxes for 'human' and 'nature' where people are separate from nature and dominant over it.

As we've seen, the idea that we inhabit a shared place in the natural world is often an ancient worldview, ousted by science. However, ideas of embodied cognition and extended mind have recently emerged, where our minds and the wider environment operate as a coupled system.[68] Such philosophical and psychological arguments firmly embed us in the environment. More and more, researchers are highlighting the integration between biology, phenomenology and the sciences of mind.[69] Humans are biological beings evolved to make sense of the natural world we're embedded within.

A little earlier than phenomenology, the science of ecology also focused on relationships, examining all the self-organising systems within systems that form the natural world. It explores a complex network of nature where one change can have an unexpectedly large effect elsewhere. Simple rules of cause and effect don't apply. Yet, in science, we ask questions that require studies to control variables to isolate causality. A systems view is more inclusive of the relationships within the natural world and helps understand the complex nature of reality. Relationships matter. Nature is embedded, interdependent and dynamic. Fixed laws do not apply. And that includes humans – a reality at a distance from the dominant metaphors that form our thinking today.

During the second half of the twentieth century, relational thinking developed in areas such as cybernetics, where humans are embodied as part of a living system where one part can never control the whole. Gregory Bateson (1904–80), the anthropologist, social scientist and systems theorist, wrote that we should not be working towards control based on our imperfect understanding of the natural world. Instead, we should think about how we can work towards improving reconnection with nature and the wider ecology. Before the current environmental movement developed, Bateson argued that ecological destruction was being caused by our conscious purpose and our ways of knowing and seeing the world, which separate us from nature, writing, 'We are not outside the ecology for which we plan'.[70] To Bateson, wildlife and plants were systems through which matter continually passed. Those systems did not have boundaries, but rather interfaces, such that ultimately all is one. Bateson felt that relationships should be at the centre of our epistemology and ontology. As we've noted, this is at odds with the dominant worldview, and Bateson saw remedies for this flawed way of thinking in the arts, in aesthetics and in contact with the natural world.[71]

The dualistic perspective of the separated 'human' and 'nature' can be seen as a defining feature of modernity, and it is clear that modern humans have caused damage to the natural world. Nature would be in a different state without us. The reality is that humanity exists embedded within nature. Great thinkers such as Einstein knew this, and modern scientists such as Sir Bob Watson, lead of the IPBES global assessment, talk of the need to tune into and form a closer relationship with nature to address the existential crises of climate warming and biodiversity loss.[72] Yet that human relationship with the rest of nature is not a widespread discipline and is a surprisingly limited topic of study. When the human–nature relationship with nature is broken to the extent that civilisation as we know it is under threat, what could be more important?

At our great seats of learning, degree courses generally fall into studying humans, their activities, processes and technologies, or nature. Approximately 60% of undergraduates at Oxbridge study humans in a range of topics from medicine, classics, psychology, philosophy, history and language to human activities such as law and economics. Human technologies such as computer science and engineering account for a further 20%. The remaining 20% seek to further their understanding of the natural world in subjects such as biology, geology, physics and natural sciences. These courses tend to focus on humans or nature in a dualistic fashion. Of course, others have highlighted modern fractured thinking: 'a single, underlying fault upon which the entire edifice of Western thought and science has been built – namely that which separates the "two worlds" of humanity and nature', as described by Tim Ingold. This is a fracture that divides disciplines, with those related to the human and its products on one side and the natural world on the other.[73]

Within these degree courses, there are modules where that essential relationship between people and the rest of nature may be considered. They occur, for example, within anthropology, philosophy and geography, but these are very wide disciplines within themselves. There's also an environmental branch of psychology, which may well focus on the built environment rather than the human–nature relationship. Either way, environmental psychology is not a core undergraduate topic in psychology. Similarly, environmental history is rarely a focus of history programmes. When relationships are studied, they are those between countries, and within modern societies. The question of 'what it is to be human' tends to focus on social and cultural relationships rather than the relationship with the natural world.

The fracture between 'human' and 'nature' that defines our modern world can even be detected in the solutions proposed for the environmental crisis. For instance, the UN Sustainable Development Goals can be seen as dualistic. Some focus on humans, such as ensuring health and education or ending poverty and hunger. Others focus on nature, such as conserving marine and terrestrial ecosystems. However, none of the seventeen focus specifically on improving the relationship between humans and the rest of the natural world.[74]

In sum, a revolution in the way we think released a revolution in science and industry, and a revolution in the exploitation of natural resources. Humans broke free of the bonds of nature, creating a fracture so profound that its separation seeps into our everyday language. With a box created for humans and another created for nature, it became unnatural to think in terms of relationships. Humans celebrated their greatness and became addicted to individualism. Nature was diminished. At its best, the dualistic scientific approach has delivered many great advances and benefits. Yet the illusion of separation hid ongoing relationships.

All the time, the natural world has been reacting on a global scale. Humanity is finding it was never in control, that it is part of a relationship dynamic. The forces unleashed by pumping one part of the natural system (coal) into another (the atmosphere) could be disastrous for the civilisation our technological advances built. Similarly, the destruction and pollution of the natural habitat is breaking the ecosystem that all life depends upon.

A question now is whether further science and technological development can and should be used in an attempt to control and rebalance the complex systems of nature. Or do we reduce control and find solutions through living more harmoniously with nature? As we will see, there is a potential harmony to be found that could bring forth meaning in life, the meaning originally promised by the technological progress and wealth that are harming the natural world. The difficulty is that the fracture is so ingrained that it constrains solutions. So profound is the split between people and nature that we struggle to see the relationship as a tangible target for action. Western patterns of thinking are likely to centre on technological solutions that focus on the symptoms of the broken relationship – rather than the root cause itself, and the forging of a new bond between people and the rest of nature.

CHAPTER 3

The Technological Ape

Technology and the human–nature relationship

Humans are nothing without their technology. We are technological apes, distinct amongst the seven great apes – formed by technology, not nature. As described by Timothy Taylor in *The Artificial Ape*,[1] we could no longer survive without it. We are born helpless and become bound to the technological culture we learn to survive in. Charles Eisenstein captures this well, noting that the technology that surrounds us helps maintain the illusion of separation from nature and that 'technology is both a cause and result of our separation from and objectification of nature'.[2]

As technology advances, nature becomes more distant. Taylor tells the story of Chekhov's encounters with the Gilyak hunter-gatherers of Sakhalin Island. How they did not understand the purpose of roads, preferring laboured travel through neighbouring marshland. To the Gilyak, travel wasn't about a journey from somewhere to someplace else. 'They were always somewhere … the road was nowhere … an obstacle … insulated from the sound and feel of the actual earth'. As it progresses, our technology tends to render the natural world mute and removes the need for it to be experienced. And so it continues: where once we foraged, we then ploughed, then built the shop, and now order from the shop without leaving our shelter, experiencing and interacting through a virtual environment. Simple everyday opportunities to notice nature diminish as nature itself declines. Taylor notes that this is not a fundamental change; humans have been creating and living within an ever more complex artificial environment for two million years.

That artificial environment includes the technology of writing. Written documents being accepted as binding have allowed the creation of 'imagined realities'.[3] Writing has allowed abstract concepts such as consumer brands to become as real as an oak tree – and more meaningful. The written word makes abstract notions real. Writing allowed companies to have the same rights as persons through corporate personhood, creating a dual reality that allowed strangers to cooperate and do business, often exploiting nature to ever greater effect. These realities can be seen as competition for the reality of nature, but can the written word also distance us from nature? In *Becoming Animal*, philosopher and cultural ecologist David Abram discusses how our ancestors were taught orally within the landscape. As the written word became the source of knowledge, these oral traditions were lost, and the natural landscape fell mute.[4]

The idea that technology is a key factor in the human–nature dynamic is supported by research. Another sign of our broken relationship with the natural world is revealed in our writing – the use of nature words in works of popular culture. Books, films and songs reflect the social context of their time. So the frequency of nature words offers an insight into changes in culture over the years. In an analysis of English works throughout the twentieth century, researchers Selin and Pelin Kesebir found a cultural shift away from nature starting in the 1950s.[5] They argue that nature references reflect their creators' minds and the social scene; the author will want the work to resonate with their audience. They considered a nature lexicon of 186 words and compared it to a comparison set of human-built-environment words. The frequency of nature words in works of fiction rose slightly from 1900 through to the 1940s, but there was a sharp decline from the 1950s through to 2000. The data shows a pause in the 1970s, with a clear decrease from 1980 onwards. Their analysis of film scripts also showed a decline from 1930 through to 2010. An analysis of song lyrics from 1950 showed a scattered range of nature word use into the 1970s, but a noticeable reduction from 1980 to 2000. The overall trend was clear – nature features much less in popular culture today than it did around 1950. As suggested earlier, language not only reflects culture, but also shapes it.

In further analysis to establish reasons for this decline in the usage of nature words, Kesebir and Kesebir argue that the phenomenon is not mirrored in rates of increasing urbanisation. However, the dawns of new technology do mirror the declines – from television in the 1950s to video games in the 1980s. Did these technologies grab attention away from nature? It seems that the technology that defines us as humans shapes us more and more.

Smartphones are a great example of how technology shapes us. Smartphones have rapidly become widespread and an integral part of our lives; by 2020, 87% of adults in the UK owned a smartphone.[6] To explore the idea that technology plays a part in the growing human disconnection with nature, our research group set out to examine the relationship between smartphone use and people's relationship with the natural world.[7] We found that higher smartphone use was related to the number of selfies taken and that a weaker relationship with nature was linked to higher levels of selfie-taking and smartphone usage times. Those who had a close relationship with nature used their phones half as much each day, took 90% fewer selfies and took 300% more pictures of nature than those with a weaker relationship with nature. That's quite a difference! No surprise, then, that people with higher smartphone use had a significantly weaker relationship with nature. This work was correlational, so it does not tell us the direction of the relationship: whether a lower connection with nature causes greater smartphone reliance, or vice versa. However, it does show there is a link between technology and the human–nature relationship.

Smartphones and selfies also return us to the humanist focus on the individual. With smartphone technology, a device designed to take and share a picture of oneself is in the pockets of 87% of people at a time when individualistic phrases have been rising for several decades. With ideal timing, the tool for selfie-taking arrived. Add the advent of social media, and selfies provide a self-presentation tool that can reflect an individual's personality and, with the ease of multiple and easy-to-edit pictures, their ideal self-concept. Selfie-taking is a good example of how technology shapes and defines human behaviours. However, fixing the broken relationship with nature cannot be about demonising technology or going back to 'halcyon days'. For the technological ape, a re-connection with nature must be part of a modern, increasingly urban lifestyle where new technology is used to *engage* people with nature. Or, as we'll consider later, perhaps technology will start to actively manage the human–nature relationship.

Fear of nature

Humans can control the artificial environments they create and feel safe within them. At the same time, technology can also be used to bring about fear. Earlier, we considered Kellert's dimensions of biophilia and that four of these were unrelated to a close relation-ship with nature. These were dominion over nature, utilitarian use of nature, a scientific relationship with nature, and fear of nature. We've seen above how our Western dualistic thinking underpinned the scientific worldview, leading to the abstract study, control and use of nature. However, we've yet to consider fear. The natural world – its extreme weather, wildlife and plants – can kill. Harari tells a story of humans being marginal and weak creatures, living in constant fear of predators for two million years.[8] Clearly, humans are physically weaker than many animals. If we were teleported to a band of hunter-gatherers one million years ago, I'd imagine many of us would feel constant fear of the threats lurking in the forest or prowling the savannah. However, a hunter-gatherer teleported in the opposite direction would probably be in constant fear of our modern urban environments too. Our roads are busy with machines that can kill, but rarely when one is sensible. It seems likely that ancient humans were cautious, attuned to threats, but well used to living with them and knowing when to take flight.

The legacy of those millions of years embedded in nature can still be seen today in responses to threats such as snakes and spiders.[9] Laboratory studies have shown quicker search ability for fear-relevant stimuli like these – with further work based on Snake Detection Theory showing that snakes, a threat to primates including humans, over evolutionary time, were more effectively detected than spiders. However, similar research has found that guns also provoke a quicker response in people, suggesting a threat response based on emotions rather than evolution.[10]

If guns can provoke a speedy threat response in UK adults, we might expect that cultural products such as books and films can also impact the way our brains are wired, and this has been the case for our perception of natural environments such as woodland.[11] We've considered the power of language, the written word and metaphors, and research has found that stories can also lead to negative perceptions of wildlife – language creates a reality.[12] In stories, wolves are generally negative, portrayed as sinister, wicked, threatening and uncontrollable creatures.[13] These stereotypes of the 'Big Bad Wolf' are 'psychologically real', informing sentiments and beliefs.[14] It has been found that such negative, and incorrect, portrayals can significantly affect levels of fear of wolves in children – this effect being stronger where wolf populations were lower, potentially hampering reintroduction projects. So negative portrayals of nature matter, but greater exposure to nature can counter such fears and bring a more positive image.[15]

Fairy tales heard in childhood can resurface in adulthood. These can be protective and spiritual or scary, fear-inducing myths such as 'Little Red Riding Hood', which is a set topic for primary school children in the UK.[16] In Western Europe, woodland myths are deeply entrenched in the cultural imagination.[17] 'Exhilarating fear' may be used to 'enhance' woodland visits, but fairy tales can affect perceptions of danger in woodland to the extent that children become afraid of these areas.[18] Little Red Riding Hood has a long history, with several interpretations and versions. It is perhaps taught in primary school for the message of 'don't talk to strangers',[19] but collateral and unintended effects should be considered. As we've seen from exploring the power of metaphors and the written word, stories about nature matter. They contribute imperceptibly to children's worldview. Given the potential negative impact on perceptions of nature at a time of a failing relationship, there are probably better ways to get across the message of 'stranger danger'. If this traditional tale is essential learning, its interpretations and wider impact on perceptions of the natural world could be taught at a later stage, alongside ensuring children are also enjoying positive information about and experiences in nature.

It's not just about fear. The power of negative cultural associations can also impact the positive emotions brought by interactions with the natural world. Some bird species are more disliked than others, with gulls, pigeons, crows, magpies and starlings suffering. Such birds can be seen as less visually appealing, raucous and harsh, vermin-like, or as 'rats with wings'.[20] Crows sit alongside wolves and spiders, with numerous accounts of connections to death.[21] Researchers have found that these 'bad animals' are seen to breach the boundaries, 'contaminate' human territory and threaten social order, challenging our dominion over nature.[22] The negative image of seagulls can be seen in the popular press with reports of a 'seagull apocalypse', and people have been known to beat them to death on the beach.[23] Recently our research group has studied the positive emotions brought by birds in the garden. As we'll see, such positive emotions can make a real difference to mental health and our relationship with nature. However, we also found that the woodpigeon brought 32% less joy than a robin. This pattern was repeated, with the smaller birds such as blue tit, robin and goldfinch bringing significantly more joy than corvids and woodpigeons. Once a weak ape with valid fears, today's dominant technological ape creates fears that steal the joy of nature.

Our pets

We've considered a long history of the human–nature relationship and why it can be said to be broken – from large-scale destruction of habitat to how the simple stories we tell our children could be contributing subtly but incrementally to that failing relationship. Clearly, industrial-scale use and control of nature are primary issues, but our actions as individuals matter too. From brushing our teeth with toothpaste that contains palm oil to spraying weedkillers containing chemicals promoted as 'garden essentials', the products we buy, our food and even our morning coffee can bring the extinction of wildlife a tiny step closer. Consumption is the main driver of extinction risk.[24] Our everyday actions contribute to the decline of nature. That includes our presence in nature, with wildlife doing less well near leisure facilities and walking trails.[25] Alongside this evidence and wider background of use and control, a right to roam can be seen as rather human-centred and dominionistic. Nature needs space to recover. Rather than a right to roam a barren landscape, I would rather have a right to enjoy more limited access to a vibrant natural world.

Unfortunately, with the close bonds to nature removed and many places lacking wildlife, the impact of our lives and actions on the natural world may rarely enter our thoughts. One interesting example is our pets. In the UK, the number of cats and dogs increased by 20% during the pandemic. There are now thought to be around 12 million of each in 17 million households.[26] Most new pandemic pets were dogs, with over half of new owners aged 16 to 34, and social isolation and loneliness being a key motivation for many.[27] One in five people who became dog owners during the pandemic, however, are now considering rehoming. Maybe something to do with the three-fold increase in hospital admissions for dog bites.[28]

Twenty-four million cats and dogs are bad for wildlife. Globally, dogs are often the most abundant carnivore in human-populated landscapes.[29] A review of 69 studies into the impact of dogs on wildlife found evidence of predation and disease transmission from the estimated 700 million dogs globally, contributing to 11 species extinctions.[30] Much of that research was into free-roaming dogs, but even dog walking can have an impact.[31] An experimental manipulation of dog walking in 90 urban-fringe woodlands found dog walking reduced bird numbers by 41% and bird diversity by 35%.[32] This was with dogs restrained on leads and in areas where dogs are prohibited or commonly walked, yet the National Farmers Union found most dogs are walked without a lead.[33] Some of the disturbance from dog walking includes the presence of a human, but further empirical research shows that birds perceive dogs as posing a greater threat than humans. Birds leave their nests earlier and stay away longer when approached by a person with a dog rather than a person alone. Such disruption increases the chances of predation and thermal stress, which can impact hatching success.[34] Yet, just as natural spaces are disrespected through littering, 'dogs on leads' signs are often ignored or pulled down at nature reserves despite the evidence that the presence of dogs can harm wildlife.

Some people have a deep lack of respect for natural environments; would they litter or exercise a dog off a lead in a churchyard?

However, cats probably do more damage than dogs,[35] particularly in urban and domestic settings. In the United States, it is estimated that domestic cats kill billions of birds and mammals each year.[36] In the UK, research studies estimate that each year cats kill over 200 million mammals, birds, reptiles and amphibians.[37] This may seem high, but with around 90% of cats bringing prey home, that would be a couple of kills a month across 12 million cats. Some argue that some of the birds caught will be weak and there are many other factors such as habitat loss, chemicals and lack of food, but all indicate a failing relationship with nature.

Over 20 million much-loved cats and dogs are killing or disturbing millions of birds and small mammals each week in the UK. When considered at scale, our simple domestic rituals with our pets can be seen as a war on wildlife. Around 250,000 stray cats roam urban areas;[38] each morning or night, millions more effective feline hunters are released into our neighbourhoods. Regular dog patrols head out into the fields, woodlands and beaches. For wildlife, this is something akin to a post-apocalyptic movie. Beyond the suburbs and globally, this is even more apocalyptic. Domestic activity is combined with industrial-scale actions such as deforestation and pesticide spraying. For wildlife, humanity is akin to Skynet in the Terminator movie franchise. Decade after decade, the war on wildlife continues.

If that wasn't enough, there are further costs from our pets beyond direct predation and wildlife disturbance. Cats and dogs have to eat, about 20% as much as humans do. Globally, pet food production is estimated to use land double the size of the UK, 5 to 11 km^3 of water,[39] plus biocides, phosphate and fossil fuel. In the USA, cat and dog feeding releases around 64 million tons of greenhouse gases each year,[40] equivalent to the annual emissions of 13 million cars,[41] with the carbon dioxide emissions of dogs being twice that estimated for cats.[42] There is some debate about the calculation of such figures, but it has been estimated that the global warming potential of dog and cat food production is twice that of truck emissions,[43] with 106 million tonnes of carbon dioxide emitted, which is more than most countries.[44] Yet pet ownership grows; the sight of a new kitten or puppy often brings joy, whereas the ownership of a greenhouse-gas-emitting car is increasingly questioned. The emissions from catering for our pets get little attention. Rather, pets, in particular dogs, are increasingly catered for with their own television channel and a warm welcome into offices or aircraft for the purposes of emotional support.[45] Some people elevate the status of their pets to 'surrogate humans'.[46] Clearly, for many people, pets are an important part of their lives; dogs and people have had a close relationship for over 15,000 years. Pets provide meaningful interaction, entertainment and companionship – but as we will see, nature also helps manage our emotions, without the ecological pawprint.[47] There are times when we could turn to nature first.

Considering the status given to pets, it's no surprise that people don't generally see pets as part of nature. Our research has found that when people are asked to write about the good things in nature, they don't tend to write about pets. Indeed, just 1% of the thousand sentences gathered referred to dogs, compared to 22% about trees, with a whole host of natural elements appearing more often, such as birds, rain or flowers. Farm animals such as sheep and cows appeared only once or twice. Despite some participants getting bogged down in definitions, people tend to have a clear idea of what nature is.

Pets aren't seen as nature, but might they play a role in developing a close relationship with it? There's been very little work on this, and the results are mixed. In a small sample of psychology students, pet owners were found to have a 7% stronger relationship with nature than non-pet owners.[48] However, this difference could come from simply spending more time outside, as walking a dog is the most frequent green-space activity in England, accounting for over 44% of all visits.[49] Or it could be that those with a closer relationship with nature got a pet to keep them company during the time they already spent outdoors. More recently, in a study of a couple of hundred adults in the UK, no link between pet ownership, as a child or adult, and a relationship with nature was found.[50] Finally, in a focus-group study, people felt that pet ownership was not a route to a closer relationship with nature, as domesticated animals were no longer natural, indeed being viewed as more human, citing aspects such as eating tinned food.[51] There are also mixed results on whether pet owners exhibit greater pro-nature behaviours; some have found they do, others that they do not.[52] Whatever the answer, it seems the impact of pets on wildlife isn't outweighed by the benefits to pro-nature behaviours or the human–nature relationship. Alongside all our other choices and activities, the impact of pets on wildlife should be considered more often.

Models of wellbeing

Whereas the wellbeing benefits of pets are often recognised, the signs of our broken relationship can also be seen where nature isn't harmed, and instead its importance is overlooked. As we'll see later, nature is good for our wellbeing, yet it is ignored in our most prominent models, such as the Five Ways to Wellbeing: connect, take notice, give, be active and learn. In the UK, these Five Ways provide straightforward and popular guidance on the steps people can take to look after their wellbeing. They're often used as a design framework, but nature is absent from the advice associated with each.

The UK Government's Foresight Programme commissioned an extensive review to produce five actions that could promote public understanding of the factors associated with wellbeing.[53] However, Foresight's 'Mental capital and wellbeing' project final report includes no specific nature-based solutions in its 317 pages.[54] There are just two sentences noting that access to green spaces contributes to wellbeing. Although the original New Economic Foundation (NEF) guidance at least mentioned 'noticing the changing seasons', many interpretations of the guidance, such as that by the NHS, overlook nature completely.[55] This is despite the fact that it's very straightforward to integrate nature into the Five Ways to Wellbeing guidance – for example:

- Connect – Social relationships are important for wellbeing. Be with and talk to people – about anything, about nature! When alone, find a friend in nature.

- Take notice – Be aware of the world around you, savour the moment, notice nature.
- Give – Take part in community life; do something for a friend – do something for nature.
- Be active – Walk or cycle when you can, to green spaces to connect with others, to notice, to give and learn.
- Learn – Try something new, rediscover your childhood wonder for nature, learn that people are part of the wider natural world and nature matters for human health.

In addition to reflecting our broken relationship with nature, omitting nature from guidance to promote public understanding of wellbeing is a missed opportunity to promote the value of, and our essential place within, the natural world. Given this omission, public awareness of the factors associated with everyday wellbeing is unlikely to include nature, which may be a factor in people seeking alternative approaches such as drinking alcohol.[56] It's therefore of little surprise that a subsequent NEF report on the application of the Five Ways to Wellbeing provides very little on the role of nature-based solutions. Instead, there's a brief mention in the planning policy of 'opportunities to traverse green space' influencing activity levels.[57] As we'll see, green space is not just an environment to cross; it's a source of wellbeing.

Despite the increasing recognition of the health and wellbeing benefits we get from the natural world, nature is missing in many models of workplace wellbeing and models of health. Our established and dominant health models are dualistic; they view people as separable from their environment, and the 'biomedical' model sees ill-health as deviation from 'normal' within the individual. The extended 'biopsychosocial' model from the 1970s reflects a 'mind–body connection' where biological, psychological and social factors play significant roles in health. Even commissions on global mental health and sustainable development focus on the environmental threats from climate change, rather than nature's contribution to mental health.[58] Models of health that view people as separable from their environment mean nature's contribution to health is overlooked. Despite it never being questioned that fish need the river, birds the sky and apes the forest, decades of research funding has been spent trying to provide evidence that people need nature. It is estimated that providing a basic level of access to green spaces, a rather basic form of habitat, could prevent over 40,000 deaths each year in cities across Europe.[59] The power of dualistic modernity that separates humans and nature is such that we have to prove our natural habitat is good for us.

The battle for attention

With nature ignored by our models of health, disappearing from our landscapes, fading from our books, controlled, exploited, dissected and feared, it is no wonder we are turning away from it. Working with the National Trust, we surveyed the UK population and found that 80% of people rarely or never watched wildlife, smelled wildflowers or photographed nature. A total 62% of people rarely or never listened to birdsong or took a moment to notice butterflies or bees. Just 6% celebrated natural events such as the longest day of the year. Nature does not capture most people's attention, but in many ways, we can ask: why would it?

As we've seen, affluence is built on a consumer economy that meets needs created by the market and technology. Abstract consumer brands have become as real as the trees, and they battle for our attention. Attention has become a global currency, and businesses spend billions competing for it; over £20 billion a year is spent on advertising in the UK.[60] When a company spends a million to grab attention for brand X away from brand Y, nature is on the sidelines – quite literally at times, in the cracks beneath the billboards. Nature does not have a marketing budget. Nature has no new styles; this year's robin is the same as the previous years. A 0.5% levy on advertising could give nature an advertising budget of £100 million: a little less than the amount spent on advertising by the UK Government in response to the pandemic in 2020.[61] We face an emergency on the state of the natural world; why not alert people to it and bring their attention to the benefits of nature?

We have seen how nature comes to mean less in our lives with each new wave of technology. Novel technology shapes our culture, our thoughts and our actions. Technology creates new things we've come to need in our lives; we've gained happiness from radio, television, home computers, video games, the Internet and smartphones. Choices and options multiply, life becomes fast-paced and complex, attention spans shorten.[62] Until, that was, the restrictions to manage a pandemic forced us to pause and stay close to home. During the spring 2020 lockdown, people visited and noticed nature more. In the UK, visits increased by 40% and noticing nature increased by 74%, improving people's relationship with nature. This is important as in our analysis, we found that the increases in noticing nature were associated with a greater sense of having a worthwhile life and more pro-nature conservation behaviours.[63] Recent nature visits, on the other hand, didn't account for higher levels of wellbeing or pro-nature actions. However, as mentioned earlier, this new-found relationship with nature didn't last; a year later, in spring 2021, levels of nature connectedness dropped back by 25%. It wasn't that people had stopped visiting nature – visits remained high, outdoor places provided a safer environment, and as we've seen, there were several million more dogs to walk. But as lockdowns eased, people took less time to notice nature. Their attention was diverted elsewhere.

Nature isn't packaged into consumable bites and has returned to the background for many. So, when nature does come to our attention, it needs to be in a positive way. The potential for stories and negative seagull press coverage to instil a fear of nature may seem trivial. Yet corporations know the power of emotions, how they drive attention, engagement and participation, and trump facts.[64] Social media uses 'emotional architecture' to allow users to express emotions; emojis are part of a carefully crafted, positively biased product designed to capture attention.[65] Similarly, emotions play a crucial role in advertising.[66] Humans are emotional and irrational creatures.[67] In 1928, Edward Bernays created modern public relations by targeting people's emotions to manipulate them to purchase items that they did not need.[68] Bernays used emotions to mould public desire and created a consumer culture, shifting social norms at a societal scale.

These ideas form the foundation of modern consumerism and increased self-absorption in Western society. We are consumers.[69]

Modern western culture is exemplified by consumerism and individualism, an all-consuming self. Therefore, the goal of marketing can reach beyond the purchasing of things to the improvement of the person, through making people dissatisfied with who they are and what they have. Richard Eckersley argues that much of the consumption that goes beyond meeting basic needs is 'morally located' within self-centredness, envy, avarice, lust and anger – traits traditionally seen as vices. Consumerism has little room for patience, moderation, prudence and compassion.[70] Even if we can see beyond ourselves, our choices are many, and our world is increasingly complex and fast-paced. When a mundane pause occurs, such as waiting at the bus stop, advertising and technology rather than nature win the battle for attention.

There are signs, though, that a desire for experiences is replacing the pursuit of possessions.[71] With attitudes towards capitalism changing, for example, it's claimed that 75% of younger people in the UK believe the climate emergency is a capitalist problem.[72] Yet corporations will be quick to spot these trends, in order to be successful in the battle for attention and monetise them.[73] Furthermore, perhaps these attitudes don't transfer into behaviours, with other surveys finding that more than half of younger adults in the UK buy most of their clothes from fast-fashion brands and out-consume older adults, yet they care more about sustainability.[74] This highlights the complexity of human behaviours; the broken relationship with nature is not about generations – as we've seen, there is a long history and dominant worldview that affects us all, stealing nature from our lives. If young people care more about climate and sustainability, then to address the destruction of biodiversity, that care needs to continue into overcoming centuries of dualistic outlook to form lives within societies that are interconnected with the natural world.

The human–nature relationship is broken. More than most, the UK and English-speaking former British colonies are not nations of nature lovers. Whether in business or at home, many people see nature as something to use or to control; these perspectives on nature dominate. Although these relationships were once fundamental to our survival in the natural world, 'progress' and technology has allowed the damaging relationships to be industrialised. And as the exploitation of nature has increased, people have been placed outside it, within a system seeking its conquest. Nature's beauty is a romantic ideal: appreciated, but ultimately not valued as our habitat became domestic. Technology and progress have delivered wonderful things: unimaginable standards of living, eradication of some diseases and lives lived long, but increasingly lives lived unsustainably and devoid of nature. Those with a close connection with nature are often depicted as tree or bunny huggers, sitting cross-legged in a 'soft-focus' world – appreciating nature is presented as a pursuit for those equipped with a mac and wellies. Yet being and feeling part of the natural world, finding meaning and value in nature, is not a search for transcendence and a superior state. Reconnection is simply about being a human within the rest of nature.

CHAPTER 4

Hidden Connections with Nature

Everything is connected

We've seen plenty of evidence of our broken relationship with nature, and have considered how our modern worldview and ways of living have led us to this point. How our sense of embeddedness in nature was lost. Our broken connection with the natural world is a tragic and thoroughly depressing story. We need some hope! Before introducing the psychological concept of 'nature connectedness', let's get physical. Let's blur them together and get magical, because there *is* hope in our latent connections to nature. There are exciting, wondrous stories from research that suggest we still possess imperceptible 'physical' connections to nature. But, of course, if we've learnt anything from the story of our broken relationship, it is that everything is related. We'll start to see how the physical becomes psychological; there isn't a neat boundary between the two.

There's a great deal of evidence on how being near to nature is good for human wellbeing. Simply being close to nature is linked to wellbeing and health benefits such as reduced blood pressure, respiratory and cardiovascular illnesses, and positive side effects such as improved pro-social behaviour.[1] On the wellbeing side, even a simple view of nature can be of benefit and exposure to nature helps restore our emotional and cognitive resources.[2] These health and wellbeing benefits are signs of latent connections. They're wonderful, and many others have focused on nature as a cure and route to restoration, but nature is not a resource, even if it's for human wellbeing. Sure, it's great to know how exposure to nature benefits us, but it is not a magic pill. We can demean it by thinking of it in terms of 'a dose' taken for our restoration. Nature's story is one of relationships. Nature should not have a 'part-time role' in our wellbeing. A full-time relationship is needed for a sustainable future and optimal wellness.[3] In our dualistic world, a nature for health perspective is an easy path to follow – one I strayed down for a little while and will again, given the ingrained Western thinking we live within, and because research funding and journals tend to focus on human health. Embedded as we are within this worldview, it's easy to forget that nothing is separate from anything else.

Despite the commonly misrepresented notion of 'survival of the fittest', it is clear that in nature, everything is connected and there are positive relationships between species. Life comes from unity and sharing. Peter Wohlleben tells some of these fascinating stories in *The Secret Network of Nature* – how within the entangled web of nature, strange relationships exist such as those between wolves, bears and fish; between deer and trees; and between ants and aphids. The complexity of the ecosystem is such that simple rules of cause and effect don't apply. Given these stories, it should be no surprise that people are involved in these relationships too. Let's consider a few that apply to humanity. We'll start with a simple wider story, about how people are good at spotting biodiversity. In a research project, we asked people to rate the amount of wildlife in various locations. Experts then visited those locations and did an ecological survey of each. We found that the general public's ratings of perceived biodiversity related very well with actual biodiversity recorded by experts.[4] This innate ability to recognise differences in biodiversity has been found by others across various habitats,[5] and suggests a latent connection to the secret network of nature.

Connections within us

From that wider scale, let's zoom in, because all forms of life are connected. In recent years research into our microbiome – that's the huge array of microorganisms that live on and inside us – has shown that it plays a vital role in our health and wellbeing. For instance, research into the 'microbiome–gut–brain axis' has revealed the impact of the gut microbiome on mental health conditions such as anxiety and depression.[6] The human gut contains around 1,000 species of bacteria and is one of the Earth's densest microbial habitats, home to trillions of microorganisms such as bacteria, fungi and viruses.[7] Our intestinal microbes influence our cognition and mood. Let's consider some of the research in this area, work that highlights the entanglement of humans and nature further still.

First, thinking about how our microbiome influences mental health, Grace Lucas discusses how the dominant models of mental health focus on the head, with mental disorders being seen as brain disorders.[8] We've seen the importance of metaphors, and we regularly see images of the brain used to depict mental health. However, this recent research shows that mental wellbeing is not simply located in the head but is also related to the body. It also demonstrates a mind–body–nature relationship. The microbiome-gut–brain axis cuts through the dominant Cartesian boundaries that separate mind and body. However, this doesn't provide a simple whole-body location of mental health. These microorganisms are part of wider nature, and these findings challenge us to rethink what it is to be human. We are not distinct and bounded individual entities. Just like other animals, we are a product of continual interactions with microorganisms and the wider natural world.[9] Microorganisms are not simply pathogens that threaten health. Our being depends upon these third-party organisms. Each person is a community of half human and half microbial cells.[10]

Jake Robinson, an ecologist and researcher fizzing with ideas and enthusiasm, tells us how humans are walking ecosystems. We host billions of microbial organisms in a symbiotic relationship that forms an ecological unit.[11] That community of beings goes beyond the body, as all forms of life are connected.[12] It's even hypothesised that the gut–brain axis in animals is an evolutionary analogue of the root–leaf axis in plants, linking animals and plants together.[13] Whether this is borne out or not, the tree of life is more a web of life with both cross- and linear evolution, such that 8% of the human genome is of microbial origin. As humans are a community of microorganisms, and those organisms play a role in our health, our everyday being amongst them is likely to matter. Handily, we inhale tens of millions of bacteria each day, as the air contains nature unseen.[14] Trees positively impact each breath, as bacterial species richness is linked to the number of trees and how close they are. Even while containing more bacterial species, urban woodland has fewer pathogenic species than urban sports fields.[15] These pathogenic species are far less common than the adverts for anti-bacterial wipes and sprays would make you believe. Those products may kill 99.9% of bacteria, but only 1% of microbes cause human disease.[16] The perception that bacteria are bad creates a fear that can become a barrier to getting outside and forming a close relationship with nature. Whereas in reality, just like the visible nature of our evolution, these invisible parts of nature are 'old friends'. Sadly, just as visible nature and wildlife is disappearing, microbial diversity such as bacteria and fungi forms a large part of the biodiversity that is being lost. Just as they are essential to our human ecosystem, these microorganisms are essential for wider ecosystems too.[17]

Tobias Rees and colleagues discuss the profound implications of humans being integrated with the natural world.[18] How it challenges the distinction between the arts, where human culture places us apart from nature, and the natural sciences that study the non-human. Life scientists and researchers in the arts must think beyond the notion that humans are more than nature. Earlier, we considered alternative philosophies to dualism, but humans being interwoven with the microbial world is not a reasoned theory about reality, such as 'I think therefore I am'; it *is* a reality. A reality where one might say, 'I am a community of human and microbial cells', or 'I think because we are – mere nature'. Centuries-old dualistic approach to thinking about reality has caught hold, providing the foundation for the destruction of large parts of the natural world. This 'enlightened' dualistic thinking has become a dogma to be escaped from for a sustainable future. What would Descartes think of this mind–body–microbial link?

How nature manages our moods

The story of how humans benefit from unseen nature perhaps helps explain another example of the latent relationship between people and nature – how the latter seems to affect the structure of our brains. Deep in the brain, the amygdala plays a crucial role in processing fearful and threatening stimuli and plays a part in how we perceive emotion in social situations.[19] Remarkably, the amount of forest within a couple of kilometres of a person's home has been found to be related to the structural integrity of this almond-shaped structure within the brain.[20] This matters, as issues with the integrity of the amygdala are linked to anxiety.[21] The forests in the study were urban in areas, woodlands with 30% canopy cover and trees over five metres tall, but outside urban parks and gardens. Although we no longer see the forest as a giving parent, even a relationship that's far from embedded makes a difference to our brains. It's a remarkable human–nature connection.

Trees are the most salient features of the everyday landscape, and we've seen how their presence and effect on the microbiome matter. What happens when we get closer and touch a tree? In a laboratory study, people placed the palm of their hand on a range of materials with their eyes closed – marble, tile, stainless steel and untreated white oak – for 90 seconds. Measures were taken to control for differing textures and temperatures and the participants were connected to various sensors. It was found that touching the oak produced physiological changes in the body[22] – namely, calming the prefrontal cortex and greater parasympathetic nervous activity, indicating physiological relaxation. This same effect was found in a second study when comparing untreated to treated wood.[23] A similar response has been found when people spend time in woodland.[24]

Even indirect contact with nature can elicit these rather magical effects. Simply viewing an image of roses for three minutes has been found to calm the prefrontal cortex and produce a positive physiological response, this time a reduction in the sympathetic nervous activity associated with threat and anxiety.[25] Returning to the amygdala, research has used functional magnetic resonance imaging (fMRI) to scan activity in the brain. This has shown how viewing a single white chrysanthemum for several seconds reduces amygdala–hippocampus activation, blood pressure, stress hormones and negative emotions.[26] These responses were generated automatically, suggesting they are hardwired into our brains. Similarly, a walk in nature has been found to decrease amygdala activation, whereas it remained stable in an urban walk. This suggests nature can help prevent mental strain and potentially disease.[27] As our brains evolved to make sense of the natural world, this is perhaps no surprise. Research has found similar results even when we only think something is from nature. When people were told an ambiguous sound was a waterfall rather than a factory they reported it as more relaxing. Physiological measures showed this wasn't simply subjective bias – the researchers also found stronger 'alpha band' activity in the brain, activity that is linked to calmness.[28] Again, deep down our brains are biased towards nature.

Our physiological response to looking at roses, being in woodland and touching wood have a common theme and can all be explained by a simple model of human emotional regulation or, more simply, how we manage our moods. This story also shows a journey from nature, through the senses and nervous system, to the heart and brain – blurring the divide between the physical and psychological, the body and mind.

It's important to remember that emotions aren't just feelings or by-products of life useful for writing poetry. Emotions are features of our nervous system, heart and brain that are fundamental to human function, with emotion preceding thinking. So, regulating emotions is very important for health and wellbeing and the continual function of human life.[29] As different emotions come and go, they shape and direct what we do. They help us deal with everyday demands in an appropriate way. From what we've considered already, it's little surprise that the role of nature in the regulation of emotions is often overlooked,[30] despite it being clear that people seek out nature to help manage their moods.

Paul Gilbert OBE developed the three-circle emotion reg-ulation model in order to simplify complex physiological process-es.[31] This model is useful in explaining how our responses to flowers and trees regulate our emotions, to help us function and keep us well. The three-circle model represents dimensions of our emotion regulation system we experience each day – drive, calm and threat. Each circle or dimension brings different feelings – such as anxiety from threat, joy from drive or contentment from calm. Each circle also brings different motivations – such as avoidance of threat, the pursuit of joy and resting for calm. There's an interplay between the three emotions and motivations, and for wellbeing we need a balance between the three dimensions. When these emotions become unbalanced, perhaps if we're constantly driven and pressured by work or study, there can be little calm, and our threat response can become overactive. We can become anxious when simply receiving an email from the boss, for example.

We all know the feeling of anxiety, and the three circles involve the sympathetic and parasympathetic arms of the nervous system. We can return to consider our interactions with trees, flowers and wood in this context. To indicate activity in the branches of the nervous system that control the heart, studies measured heart-rate variability. They indicated that being in the woods activated the parasympathetic nervous system associated with contentment and calm. Similarly, touching oak led to greater parasympathetic nervous activity, indicating physiological relaxation. Finally, viewing roses reduced sympathetic nervous activity associated with drive and threat. In comparison, the urban control environment stimulated the sympathetic nervous system.[32] Through helping to provide balance, nature helps maintain positive emotions. This can lead to enhanced immune function,[33] therefore also providing a mechanism to explain the long-term benefits of nature exposure.

Thinking of our emotions in terms of a balanced system, a system within a dynamic human ecosystem that also requires stability, brings to light a parallel with the wider ecosystems we are part of. When natural habitats are degraded ecological stability suffers, which can lead to shifts in ecological communities, loss of wildlife and a state of ecological collapse[34] – just as a lack of balance and stability in humans can lead to anxiety and depression.

Nature makes sense

Given these parallels and that everything is connected, it's little surprise to see that the role of nature in managing our emotions doesn't stop with trees, woodlands and flowers. Higher levels of biodiversity, in general, have been linked to more positive emotions. In our research, we've found a strong relationship between levels of biodiversity and a positive emotional response to those more biodiverse spaces. People reported being happier in sites with a greater variety of birds and habitats. In addition, these relationships were stronger when people thought the site was wildlife rich, even if it was not. We found that both real and perceived biodiversity bring positive emotions.[35] Results vary, however, with some finding that perceived biodiversity matters, but others that actual biodiversity matters, explaining the restorative benefit of urban parks to a much greater extent than the park's facilities.[36] However, more recently, a large-scale study involving 26,000 people across 26 European countries found that a wide variety of bird species, rather than mammals and trees, was associated with increased life satisfaction at a similar magnitude to that linked to income. The reasons for this link were proposed to come from the diverse natural landscape required to support a variety of birds and the direct sensory experience of those birds.[37] Further work in Canada has shown that the benefits from birds seem to be unconscious, as residents did not mention birds when surveyed about neighbourhood satisfaction, even though data analysis revealed that bird diversity was a factor.[38]

It's interesting to consider how and why laying a palm on some oak or viewing some flowers leads to physiological changes within the body. People's preference for natural scenes has been linked to our evolutionary past, as the innate affective responses from flowers and trees come first; we feel before we think when sensing nature.[39] The simple sensing of nature leads to physiological changes. We respond faster to natural scenes than urban scenes, and natural images are thought to be processed more effectively than artificial stimuli, with this 'processing fluency' accompanied by positive emotions.[40] Hence, perceptual fluency is associated with autonomic nervous responses like those seen when viewing flowers or touching wood.[41] This fluency could well be based on our brain's long evolution in natural environments, such that natural forms are simpler for us to process.[42] The 'back to nature' theory suggests that as 99.99% of our evolutionary history has been based in nature, our physiological functions are best adapted to natural environments. In contrast, the modern urban environment is artificial and elicits a 'stress state' in people, with natural environments bringing us closer to our relaxed, natural state.[43] Our senses evolved to make sense of the natural world, so when engaged with nature, the various systems in the body are doing what they evolved to do; it's a good fit. In comparison, urban environments are not an ideal match.

Nature is constantly in touch with us, from the unseen within us to when we are among trees or seeing flowers and biodiversity; nature helps manage our moods. The physical becomes emotional in a state of flow. Although many may feel disconnected from nature or not even think about it, research shows that humans are deeply embedded and interconnected with the rest of the natural world. What was once our habitat, for most of human existence, is good for us – a symbiotic relationship found throughout nature. In the end we're just like fish, birds and other apes.

CHAPTER 5

Nature Connectedness

Introducing nature connectedness

We have seen evidence of the physiological connections between humans and the rest of nature, via the senses and microbial relationships. These more physical and unseen connections show that people are, often unknowingly, part of the natural world. We've also seen that the 'physical' becomes 'psychological' and that those latent connections manage our emotions. This brings us to nature connectedness, a term from psychology that captures our emotional relationship with nature and whether we feel part of nature. This chapter introduces the science of nature connectedness. What it is, how to measure it, how it varies across the population, some factors linked to it being higher or lower, and why it's an essential target for a sustainable future.

Our relationship with nature is complex, and while nature connectedness includes various aspects of that relationship, it also simplifies it – reducing it to a number for the purposes of scientific research. This is problematic, as I have argued that science and understanding by reducing things to numbers is part of the broken relationship with nature! When attempting to understand something through numbers, the measurable becomes primary and reality secondary. That is the dominant worldview: numbers make things real. As a person immersed in that worldview, I am guilty of the same by highlighting the 'reality' of nature connectedness because it can be measured. However, to communicate there is a need for a shared language, and to convince others of facts there is a need for evidence. People might intuitively accept the positive relationship between a fish and the river, but so deep is the fracture between people and nature that decision-makers often do not accept that the human–nature relationship is tangible, let alone that it matters. There is a need, then, to make the point in a common language. To elevate the reality of being human within nature to their imagined reality.

This reality is captured in my definition of nature connectedness: 'a realisation of our shared place in nature, which affects our being – how we experience the world here and now; our emotional response, beliefs, and attitudes towards nature'. 'Nature lover', although limited by simplicity, provides an accessible shortcut. When two people are in love, they realise their shared place together, and it affects their day-to-day lives, experiences and behaviour towards each other. People tend to accept that love is real, even though it is not a number.

A useful and accessible way to understand nature connectedness is through analysing the language people use to talk about nature. When Melissa Hatty and colleagues categorised people's concepts of nature, they found three themes: descriptive, normative and experiential.[1] Descriptive is language describing plants, wildlife and landscapes, and represented 73% of responses. Normative language is often used by those with nature expertise, language such as 'conservation' and 'biodiversity', so no surprise that less than 2% of people mentioned such terms. Experiential language included positive emotions, feelings and activities in nature and accounted for 3.5% of responses. The remaining people mainly used language from more than one category. The categories allowed the researchers to find that people with lower nature connectedness scores described nature in simple descriptive terms. People with higher nature connectedness scores, meanwhile, used more experiential terms, or terms from more than one group. So, it is concerning that the vast majority of people think of nature in simple descriptive terms, such as types, processes and characteristics of the various parts of nature – the labels we use. This reflects how we're schooled to think about the world, but also helps show how nature connectedness is experiential and emotional.

Having defined what nature connectedness is, let's take a brief look at its history. It is a relatively recent psychological construct grounded in scientific study and accepted internationally. Nature connectedness is the formal term; when people talk about 'nature connection' and 'connection with nature', they may or may not be talking about this carefully defined and measurable construct. That can be confusing, but they are friendlier terms and can make talking about these things more accessible – so when I use those terms, I am referring to nature connectedness.

In the year 2000 few, if any, published research papers included the term 'nature connectedness', but since then, the research evidence has grown rapidly, particularly since 2012. Two of the most important publications were published at the turn of the millennium. Firstly, Wesley Schultz discussed the connections between people and nature and proposed a psychological model often used to define nature connectedness.[2] A second paper provided a key measurement tool, the Inclusion of Nature in Self Scale. This remains the simplest and most useful method for detecting changes in nature connectedness.[3] This measure presents a series of seven pairs of overlapping circles, one labelled 'self' and the other labelled 'nature', that range from fully overlapping to no overlap. The person completing it is simply asked to select the one that 'best describes your relationship with the natural environment. How interconnected are you with nature?'

There are a dozen or so scales that measure the nature connectedness construct in slightly different ways.[4] These consist of carefully worded and tested items that people respond to. I won't go into detail, but they ask people to rate their agreement or disagreement with statements such as: 'I often feel a sense of oneness with the natural world around me', 'I feel that all inhabitants of Earth, human, and non-human, share a common "life force"', 'Like a tree can be part of a forest, I feel embedded within the broader natural world' and 'I feel very connected to all living things and the earth'.[5] Thinking back to my definition above, some items and scales are quite focused on a person's belief and thinking about their connection with nature; this is often termed a cognitive rather than emotional relationship.[6] Ultimately, the various scales all do a similar job and allow us to put a figure on an individual's level of nature connectedness. Just as scales can measure how outgoing or shy a person is, we can measure how close a person is to nature.

Nature connectedness being measurable means scientific studies can be undertaken to help understand it and improve it – to improve our relationship with nature.

Earlier in the book, we considered the birth of humanism and the rise of individualism, and how modern life is creating an epidemic of narcissism that is linked to a more distant relationship with nature. One nice example of research that helps understand nature connectedness and the human relationship with nature took place in the early days. In 2005, Cynthia Frantz and colleagues published a research paper titled 'There is no "I" in nature'.[7] This research set out to test the idea that narcissism and the separation of the self from nature are core issues in the environmental crises. To manipulate self-awareness, they simply placed people in a cubicle with or without a mirror. They found that for those who cared less about nature the mirror decreased nature connectedness, whereas the presence of the mirror made no difference to nature connectedness for those who cared more about nature. The study showed that greater self-awareness lowered nature connection. They also found that when people with higher levels of narcissistic traits of exploitation and entitlement shared a cubicle with a mirror, their nature connectedness decreased. Through being able to measure nature connectedness, the results of this research support the earlier arguments in this book that the rise of attention to the individual is a problem.

There are many clever studies on the human–nature relationship like this, enabled by the focus on nature connectedness. But let's return to some fundamentals. As the measurement scales reveal, the key point is that nature connectedness is not simply exposure to nature or living near nature, often measured by time spent in nature or visits to it. Over the decades, a large amount of research that has established the benefits of nature for human wellbeing has measured time in and visits to nature; they're both straightforward to measure, so that makes a lot of sense. Yet, as we'll see in more detail later, nature connectedness itself has benefits over and above time in nature – for both human and nature's wellbeing. Hence, nature connectedness is about a relationship rather than time. As an analogy, you might spend eight hours a day sitting close by someone at work, more time than you might spend with loved ones at home. Time can be, but is not necessarily, an indicator of a close relationship. People spend time outside in nature for many reasons, such as to exercise a dog or themselves, perhaps while wearing headphones that reduce engagement with some of the joys of the natural world. Research studying identical and fraternal twins also supports the distinction between connection and contact, as they have differing influences. Indeed, this research suggests nature connectedness is partially heritable, but that the right environment, rather than genes, explains most variation between people.[8]

As we'll see in more detail in chapters to follow, a great deal of nature connectedness research has focused on establishing the benefits of a closer relationship with nature for human wellbeing. Nature connectedness certainly brings benefits in this area. Indeed, the relationship to feeling one lives a worthwhile life has been found to be four times more important than the existing benchmark of socio-economic status.[9] Other research has found that people with higher levels of nature connectedness are more likely to do more for nature, both in reducing their impact on the environment by using fewer resources and taking positive actions to help wildlife. In both senses, nature connectedness really matters.

These benefits have seen nature connectedness called a basic psychological need. Through reviewing research evidence and comparing to criteria for other basic psychological needs, two sets of researchers have concluded that the evidence supports that categorisation. That is, humans have a fundamental need to feel a connection to nature in an emotional, cognitive and physical sense. This argument is made because the benefits to people come throughout their lifespan and largely transcend socio-economic status and cultural boundaries.[10] The importance of the construct is also illustrated by its inclusion in the Gallup World Poll, which tracks important issues worldwide.[11] Further, a focus on nature connectedness was amongst the recommendations in the evidence report produced by the Stockholm Environment Institute for the Stockholm+50 meeting on environmental action convened by the United Nations General Assembly in June 2022.[12]

Calls for reconnection with nature and a new relationship with nature have increased but have sometimes been vague, with fragmentation around what nature connection is.[13] There's also been little concrete guidance towards this goal, with some simply conflating time outdoors with nature connection. The psychological construct of nature connectedness introduced in this chapter provides a measurable focus within the fragmentation. It also provides a meaningful focus as it is based on scientific study. The concept has provided an evidence base of benefits to the wellbeing of both nature and people that place it as a basic psychological need, and one that can unite human and nature's wellbeing.

Nature connectedness across the population

We'll consider the benefits of nature connectedness in more detail later, but first, let's look into some fundamentals. Quite a lot of the early work into nature connectedness was based on small samples that didn't represent the wider population too well. A circular issue here was that large-scale representative work is expensive, and funders aren't inclined to fund research without good justification. Therefore, to establish a better evidence base in the UK, I was part of a project set up by Natural England with partners from Historic England, the National Trust, the Royal Society for the Protection of Birds, and the Wildlife Trusts. This project enabled us to ask a large representative sample about their nature connectedness, experiences in nature and wellbeing.

As we have seen, although a range of nature connectedness measures are available, they weren't suitable for use in a national survey due to length, and some of the items are rather complex and esoteric. Most measures are also designed and tested for use by adults. However, we were interested in children too, so that patterns across the lifespan could be considered. So, the project team developed a short, simple measure suitable for use with both children and adults. The new scale was the Nature Connection Index (NCI). The items in the scale considered people's emotional and experiential relationship with nature, and people's responses were scored and weighted to produce an index that ran from 0 to 100.

A few thousand people aged from 7 to 95 years completed the new index, and they were selected to be representative of the wider population. For the first time, this survey produced an average level of nature connectedness across the population, and it was 61. On its own, this means little, but a slightly closer look allows some fundamentals to be established. Women scored higher (64) than men (58) and this difference was statistically significant. Elsewhere, the differences between various groups were smaller. For example, socio-economic status was considered, with those in more professional occupations scoring 64, compared to 61 for occupations considered unskilled by those that classify such things. People that worked scored 60 compared to 61 for those not working. As suggested earlier, although there are inequalities in physical access between communities, nature connectedness transcends ethnicity, with scores for adults falling between 61 and 63 for those who completed the survey from Black, Asian and White communities. We can all find a love of nature. Although, at present, all broad communities have a failing relationship with nature, on average at least.

The teenage dip

Adolescence is a time of many changes, a time when we form our own identity and more meaningful friendships. Group acceptance matters as childhood characteristics combine with emerging adolescent traits. Plenty is going on, physically and socially. So, it's perhaps to be expected that the scores for children revealed something interesting. Although starting close to adult levels at the age of 7 years, there was a sharp dip in nature connectedness from the age of 10 or 11 and into the teenage years. This reached its lowest point at around 13 to 15 years old, with a slow recovery to the adult population mean at around 30 years old.[14] From a score of 64 aged 7 to 9 years, to 47 aged 13 to 15 years, there is a sudden 25% drop in nature connectedness. This 'teenage dip' has also been identified in other research in the UK and elsewhere, such as in China.[15] A study of close to 30,000 Canadian children also found the teenage dip and evidence for why it matters. Those children who felt a closer connection to nature had 25% fewer mental ill-health symptoms.[16] Childhood is an important time to develop a relationship with nature. Early life experience is fundamental to how children develop, and this continues into adolescence, a period of rapid growth, learning and brain development. The teenage years are a dynamic time when young lives can pivot rapidly. It is a time of opportunity for positive impact, for adolescents themselves, and for a sustainable future.[17]

It seems likely that during the complex teenage years, nature, and one's relationship with nature, may lose relevance and importance amid other meaningful relationships. Many of these relationships occur at school, which includes other changes, such as the move from primary to secondary school in the UK. This transition matches the timing of the drop in nature connectedness well. There is a swift focus at secondary school on exam grades and the 'core' academic subjects such as English, maths and science. As we've seen, science can bring about an abstraction of nature. Nature is divided into parts and labels, functional units within processes to be memorised. Outside school, children change their use and requirements of nature and natural spaces as they grow older. For instance, local woodlands explored with parents at a younger age can provide opportunities for developing a sense of identity with peers. Yet when a few teenagers gather between the trees, it can be poorly understood by the local community and landowners. Curiously, although nature remains a favourite place for teenagers, most take part in very little nature-based activity.[18]

The 'teenage dip' can seem at odds with youth climate strikes that were hugely successful in bringing urgent attention to the climate crisis. So it's fair to ask: how can today's youth be disconnected from nature when they're deeply concerned about environmental issues? There are two key factors. First, the climate strikes and nature connectedness are related but different. Second, the dip is an average; just as in every generation, there are also adolescents with a very close relationship with nature.

Looking at the first factor, Greta Thunberg's school strikes of 2019 were 'to protest against the lack of action on the climate crisis'.[19] The UK Student Climate Network's mission also focused on action against climate change.[20] Rightly so, as younger people will experience more of the consequences of a warming climate. As we've seen, nature connectedness describes how close an individual's relationship with nature is, how much they enjoy nature and its beauty, how important it is to them, whether they feel part of nature and whether they treat nature with respect. So there's a link between this and caring about the climate crisis, but some key differences can be seen by looking at the content of Greta Thunberg's powerful and effective speeches of the time. In nearly 5,000 words, the most frequent words included climate, people, crisis, emissions, children, future, countries, leaders and carbon. The words nature, wildlife and biodiversity did not appear, although there are six references to extinction.[21] The youth climate strikes focused on the threat to young people's future from climate change and the need to reduce carbon emissions. Reducing carbon emissions can be seen as a fix for the products of the human–nature relationship, treating the symptoms of our broken relationship with nature. The human–nature relationship is the root cause and wasn't the target. Nature connectedness provides a focus for tackling the causal issue and can help describe what a future with a more sustainable relationship with the natural world could look like, for both climate and biodiversity.

The second factor that helps explain the mismatch between younger people's concern for the environment and the teenage dip in nature connectedness is simply that people vary so much. The dip from a score of 64 at the age of 9 years, to 47 at 14 years is in the mean level of nature connectedness. The mean score represents the person towards the middle of a range. There are still teenagers with a very close relationship to nature. Working with the RSPB, we found around 18% of children had a strong connection with nature, compared to 46% of children who had a low connection to nature.[22] It may well be that many of those supporting climate action are part of the smaller group; sadly, in that survey, most children, like most adults, weren't actively conserving nature. Of course, the reality is that it's not about generations. Adults and children together as families consume more, and people who earn more consume more.[23] Given the global scale of the environmental crises, there's a broken relationship between *most* humans and nature.

Children's connection: families and green spaces

Although there has been research into childhood experiences in nature, there is less work focusing on the factors involved in children's nature connectedness.[24] The national data that identified the teenage dip also allowed an initial exploration of these factors.[25] The analysis of this data showed that, as in adults, girls had a closer relationship with nature than boys, but nature visits were not related to nature connectedness. This might seem surprising – but as we've noted, spending time isn't the same as a close relationship. Time is often linked to short-term increases in nature connectedness, and some research has found that simple contact with green spaces such as parks does not always lead to short-term improvements in nature connectedness.[26] It's what happens during the visits that matters. In the survey, nature visits included many types of green spaces, such as parks in and around towns and cities. We don't know how many of the children's nature visits were to fenced-off playgrounds in the local park. Or perhaps the play area near the car park. Thinking of a wider analogy, strictly, it isn't the number of visits to the hospital that makes people better; it's the treatment they receive during those visits. However, visits and time in the hospital will still correlate with recovery.

Turning to the main factor in a child's connection with nature, the only significant aspect was how connected their parents or guardians were to nature. None of the other aspects (sex, age, marital status, frequency of nature visits, watching/listening to nature programmes, and household socio-economic status) explained levels of children's nature connectedness. This relates well to recent studies with twins that have shown a partial heredity element to nature connectedness.[27]

Finally, the relationship between the local neighbourhood and children's nature connectedness was examined. Unexpectedly, we found that higher levels of neighbourhood green spaces were related to *lower* levels of nature connectedness. Curiously, neighbourhood deprivation was positively linked to children's levels of nature connectedness. Surprisingly, urbanicity was not a significant factor. There were a relatively small number of rural residents in the sample, but the research with twins also found urbanisation wasn't a factor. Sometimes our assumptions aren't borne out in the data, so let's take a closer look.

The link between higher levels of neighbourhood deprivation and children's nature connectedness is counter-intuitive, so it's worth considering these findings further. For example, children from higher-income households spend more time visiting nature, but then we've seen visits don't necessarily bring nature connectedness.[28] Moreover, others have found the opposite, with more privileged children having longer school days which they are driven to and from[29] – whereas children in more deprived neighbourhoods spend more time outdoors with friends, which helps form meaningful bonds to nature.[30] Earlier, we considered the role of technology. There's also evidence that shows that children living in more deprived areas tend to spend less time using smartphones, which is associated with greater nature connectedness in adults and children.[31]

The national survey of children also found that the presence of neighbourhood green spaces was related to lower levels of nature connectedness. The presence of green spaces doesn't mean they're being used, and green spaces are often of reduced quality, with lower biodiversity, in areas of greater deprivation.[32] As ever, this is just one study; we can never fully understand complex relationships, but that also shows the need to move beyond simple assumptions. For instance, providing a local park might not boost nature connectedness. Single research projects can't provide all the answers, and different approaches have different purposes. These large-scale surveys rely on self-reports and are 'cross-sectional', meaning they can show if variation in one factor is linked to levels of another. This is useful, but it is difficult to make causal links – though at the same time, if there's no relationship, there can be no causality. Finally, we know little about what happens during the visits to nature reported; how meaningful is this engagement with nature?

Some recent research confirms the importance of these questions and the complexity of these relationships. In a survey of close to 300 children living in a Turkish city, the link found between local green spaces and the ability to manage emotions was not direct; it was dependent on the children's level of nature connectedness. So, the benefits of local green spaces for managing moods were dependent on a child's relationship with nature. This work also found that the amount of local greenery and how much a child noticed their environment predicted levels of nature connection. The simple existence of surrounding green spaces was not enough to be of benefit. Children need local nature and a close relationship with it.[33] Evidence such as this has led to calls for nature connectedness to be a distinct goal in childhood education.[34] As we search for a sustainable future, making a close relationship with nature an aim of education would make a lot of sense.

The factors in a failing relationship with nature

The science of nature connectedness provides a measurable focus for insights into the human–nature relationship. Although this relationship is complex and can't be reduced to a single number, being able to measure it does help us understand its broken form and, as we'll see, ways to improve it. Earlier, we considered a research study that surveyed 16,000 people across 18 countries.[35] It showed that the UK and industrialised former colonies had a weaker relationship with nature. We teamed up with those researchers from the University of Exeter to take a closer look at that data.[36] We wanted to see how individual levels of nature connectedness across the various countries were associated with country-level measures that reflected topics consider earlier in this book. This allowed us to identify which factors at a national scale are linked to a failing or stronger relationship with nature – to see if some of the aspects considered in earlier chapters are borne out in the data. Whether the use and control of nature matter. Whether the way we farm and use our land matters. How urban living, technology, consumption and commerce matter. How country-level factors might be related to the human–nature relationship of individuals in those countries.

Earlier in the book, we noted that the countries in the survey with the weakest relationships with nature were those in Western Europe, and the Western offshoots of the USA, Canada and Australia. We saw how these countries had had particularly fast levels of growth, with personal income increasing up to 17-fold compared to 6-fold in Eastern Europe. Such affluence is built on a consumer economy, with the utilitarian use of nature producing prosperity and economic growth. So, in this further research, we looked to see if levels of personal income in a country are related to nature connectedness. We found that lower levels of nature connectedness are quite strongly related to higher income levels. Income reflects the individual's ability to spend on consumer goods for personal enhancement and satisfaction.[37] However, this route to satisfaction and happiness can lead to self-interest and mean that other sources of fulfilment, such as a close relationship with nature, can be overlooked.

With needs created by the market, nature takes a back seat, especially as access to consumer technologies increases.[38] To look at this, we considered smartphone ownership in each country and found a strong relationship between lower levels of nature connectedness and higher levels of smartphone use. This cross-country analysis supports the research with the same finding mentioned earlier.[39] Human progress is strongly associated with technological advances, yet that relationship with technology could undermine our relationship with the natural world that sustains human life.[40]

Technological advances saw people move to live in urban environments, and we found that countries with more people living in urban areas had lower levels of nature connectedness. Urban environments typically contain less nature, which produces a more mundane natural environment and an extinction of experience of the diverse wonders of the natural world.[41] This can result in a loss of orientation towards engaging with nature which becomes a social norm, thereby creating a feedback loop which permeates culture.[42] Recently, researchers explored the 'mental models' of urban residents – that's how people think about the real world. They found that urban residents had a more simplistic understanding of the complex interdependence between humans and nature, which limited their ability to live in harmony with nature.[43] Being disconnected from nature results in simplistic thinking about nature.

The extinction of experience and social norms reinforced through urbanisation could explain another factor in lower levels of nature connectedness: having fewer adults aged over 65 years old in the population. Our research found that a more nature-connected society has more older adults. This may reflect that older people were about when nature was more abundant and had the opportunity to develop a closer relationship with it. It also suggests that our relationship with nature could well be getting more distant over the decades. Good reason to foster greater intergenerational activity between those that remember abundant nature and those denied that joy.

The potential impact of the loss of nature to experience and connection brings us to levels of biodiversity. Biodiversity is perhaps the ultimate reflection of our level of use and control of nature. Where these types of relationships dominate, there will be less natural habitat and less wildlife. This is reflected in our analysis, with the level of biodiversity across a country being strongly linked to individual levels of nature connectedness. There's not a great deal of research that directly considers nature connectedness and biodiversity, yet related research supports this result. For instance, as we've seen, higher levels of connection to nature have been linked to the floral richness people perceive, and the abundance of birds.[44] Furthermore, as we considered earlier, people are good at spotting biodiversity when prompted. It is clear that nature connection and biodiversity are strongly related. Whether it is a weaker relationship with nature that leads to the loss of wildlife, or the loss of wildlife that damages the relationship, is a moot point as neither is beneficial. It is also very likely that it isn't that simple – there is probably a reciprocal relationship between biodiversity and nature connectedness.

Another key indicator of our dominion and control over nature that can impact biodiversity is how we use land. Thus, the amount of cultivated land was another country-level measure we compared to the population's nature connectedness. In our analysis, the relationship was in the opposite direction to that we expected. Higher levels of nature connectedness were moderately related to higher levels of cultivated land. To explain this, we considered that there are many ways of cultivating land. The measure may not represent industrial farming. Small-scale, more organic cultivation could use a larger amount of land for similar returns. Organic cultivation can have a positive impact, with higher levels of nature connectedness linked to greater levels of community farming[45] – although organic farmland can also be viewed negatively, as it can be seen as 'messy'.[46] In contrast, people can view large fields with tidy hedgerows positively, as to some these are the accepted landscapes of 'home' and the 'natural landscape'. That may be the case with pastureland – grassy fields grazed by livestock such as cows and sheep is perhaps 'green and pleasant land' to many.

Britain has far more pastureland than arable land, and some further analysis revealed that here there was a negative relationship to nature connection. Those countries with higher levels of pasture tended to have a weaker relationship with nature. These relationships confirm that land use is a key part of our everyday experience of, and emotional connections to, the landscape.[47] As we'll see, nature connectedness research has tended to focus on individuals and local interventions, but there is a need to consider land-use systems and how they provide opportunities for people to engage with nature.

Finally, in this study across countries, we included indicators of potential negative relationships with nature. Here, the risk of natural disasters could lead to a fear of nature. Plus, enjoyment of nature might be hampered by rain in wetter countries. The risk of natural disasters had little relationship with nature connectedness, but there was a weak relationship between higher levels of rainfall and lower levels of nature connectedness. It seems that climate factors outside human influence relate less strongly to levels of nature connectedness than indicators of the human–nature relationship such as income, consumer technology, land use, urbanisation and biodiversity.

Of course, correlation does not show causation, but the aim of the analysis was to show that country-level factors are related to individual relationships with nature. Causality isn't an issue when there is no relationship. We'll consider the causal evidence supporting nature connectedness in the chapters to follow. For now, the fact that several indicators tell the same plausible story brings a consistency that can allow causation to be suggested.[48]

Either way, the national-level measures of modern consumer living, namely income and smartphone ownership, had a reasonably strong relationship with individuals having a more distant relationship with nature. Then, measures indicating extinction of nature experience, such as urbanisation, also showed a negative relationship to levels of nature connectedness. Finally, the ultimate cost of utility and dominion, loss of wildlife and biodiversity, had a strong correlation with a failing human–nature relationship. Looking at country-level factors in the broken relationship supports the bigger picture set out in the earlier chapters is relevant. Reducing things down to a number helps us see that the way land is used, how people engage with that land and the impact of land use on biodiversity matters for our relationship with nature. The analysis also shows that the nature of society matters: its consumer economy, urbanisation and intergenerational activity are also relevant factors.

A human–nature relationship KPI

Hopefully, even though we've yet to consider the benefits in detail, it is becoming apparent that nature connectedness is a valuable concept, one that is defined, measurable and sitting at the interface of both human and nature's wellbeing. Measured by a single-item question, nature connectedness strongly correlates with country-level indicators related to a sustainable future. In the work across multiple countries, we also compared the simple measure of nature connectedness to composite measures of prosperity, development, sustainability and social progress – using indices such as the UN Human Development Index and Sustainability Development Goal Ranking. Interestingly, across the countries in our analysis, we found that these existing metrics all had a negative relationship to wellbeing, biodiversity and nature connectedness. Even though higher scores on these indices are intended to reflect positive outcomes related to the basic human needs of prosperity, development and sustainability, they fail to capture the bond between people and nature. As we've seen, this bond is essential for a healthy and sustainable life. Analysis across more countries is needed, but it seems possible that these indexes could well be anthropocentric, with higher scores strongly related to lower levels of biodiversity. Once further evidence on the benefits to human and nature's wellbeing is considered, nature connectedness can be seen as a simple yet powerful 'key performance indicator' (KPI) of progress towards improving human and nature's wellbeing.

So, let's pause and consider where we are at. We've considered how human-induced climate change and loss of wildlife shows that the human–nature relationship is broken. Still, there is hope in our latent, somewhat magical connections to nature. This led us to our emotional connection to nature and beliefs about our place within the natural world, and how this provides a focus to produce evidence on our broken relationship. A simple yet powerful analysis of the human–nature relationship across several countries suggests that our broken relationship is related to the self-interested affluence of our urbanised consumer-based societies that consume natural resources and reduce biodiversity. That the nature of society, its economy, urbanisation, and land use impact on the human–nature relationship will come as no surprise to many. However, this new perspective adds power to the need for transformational change for a sustainable future, and provides a direction and way of monitoring progress.

Nature connectedness reflects human and nature's wellbeing better than many other measures, which ultimately value people more than nature. Yet, of course, the loss of nature will damage human development, social progress and prosperity. Hence, there's a need to focus on the human–nature dynamic, as that broken relationship has led to the climate and biodiversity crises. But enough of broken relationships – let's get positive again and consider the research into the benefits of nature connectedness in more detail.

PART II

BENEFITS OF RECONNECTION WITH NATURE

CHAPTER 6

Good For You: Wellbeing Benefits of Reconnection

Understanding the benefits

I n the first part of this book, we considered the story of our broken relationship with nature and the hope of how everything is connected. This story included a journey from nature through our senses and nervous system to the heart, to see how nature manages our emotions, the emotions that form a close relationship with the natural world. The second part of this book considers the benefits of a close relationship with nature. It starts with a chapter that shows how a love of nature is good for us, demonstrating why a connection with nature is emerging as a fundamental part of a good life. When introducing nature connectedness, I noted its rapid growth and that much of that new research focused on human wellbeing. Numerous studies have considered the correlation between a close relationship with nature and various aspects of wellbeing – each one helping to establish that our relationship with nature matters. The benefits to wellbeing include mental health, which introduces a third area of a global crisis in addition to the warming climate and biodiversity loss. The burden of mental health issues is global, with over a billion people affected in 2016,[1] and most people suffering a mental disorder during their lifetime. Despite this, people with mental illness are largely neglected.[2] The issue is more serious in more affluent countries, so perhaps the mental health crisis is also related to our broken relationship with nature. The research in this chapter helps evaluate whether this may be the case – but first, let's start with the early days of understanding nature connectedness and the benefits that it brings.

Much of the early and innovative work into the wellbeing benefits of nature connectedness came from a group of pioneering researchers based in Canada. Notable early work includes studies that show how an emotional and active connection with nature is associated with several facets of wellbeing, namely vitality, autonomy, life satisfaction, personal growth, meaning in life and positive emotions.[3] In later work, these researchers also found that a close connection to nature predicted wellbeing over and above how generally connected people felt to family and friends.[4] Along similar lines, researchers in the UK found that having nature nearby can provide socially isolated people with an alternative way of feeling connected;[5] finding a friend in nature helps. Returning to Canada, another research team found nature connectedness was associated with meaning in life and emotional wellbeing.[6] So there were signs in these early days of the link to having a meaningful life, one that is fulfilling and relatively free from feeling helpless, fearful and depressed, where a person feels accepted and connected to others.[7]

While looking at the links to wellbeing, studies also sought to understand nature connectedness further by exploring its relationship to a range of established concepts. For example, it has been found that nature connectedness is linked to several aspects of personality – characteristics such as openness, agreeableness, conscientiousness, honesty and humility.[8] These studies suggest that people with a closer relationship with nature are more easy-going and able to contemplate the future. Such positive aspects of personality are linked to a broader range of coping mechanisms for the challenges of life.

We've seen that how people think about themselves is important for the human–nature relationship, and there's a body of work that also explores people's awareness of their own thoughts and feelings. So people who tend to reflect on their emotions and attitudes have higher levels of nature connectedness.[9] This sits alongside mindful attention: basically, living in the moment and being aware of things in a way that enhances awareness of the current experience.[10] Mindfulness and nature connectedness are often considered together. In an influential paper from 2011, our Canadian friends investigated the relationship between nature connectedness, mindfulness and various types of wellbeing.[11] As mindfulness is being aware of the present moment, the idea is that people who tend to be mindful will have an enhanced experience when in nature, so their levels of nature connectedness will be higher. Over three different studies the work found that a closer relationship with nature was linked to wellbeing and mindfulness. With that link established, much more research followed, such as the finding that providing people with mindfulness training can increase levels of nature connectedness and wellbeing during a nature walk.[12] So, moments of enhanced awareness of nature and reflecting on the contribution of that experience to oneself help build nature connectedness.

The emergence of nature connectedness and measurement tools around the turn of the millennium provided a great deal of scope for exploration. There had been plenty of research into humans and nature before, but the focus on a close emotional relationship and the ability to measure it brought new directions at a time when the costs of a failing relationship were becoming even more apparent. This resurgence has revealed a range of quite varied benefits of a close relationship with nature. One perhaps unexpected example is that nature connectedness is associated with a more positive body image, perhaps because a closer relationship with nature provides a broader context for people to appreciate their body as part of a wider ecosystem.[13] Other work has considered 'dark personalities', those people who thrive on manipulation, cynicism, self-centredness and a callous disregard for others, sometimes labelled as sadism or psychopathy. This cynicism, callousness, lack of empathy and emotional connection to others was associated with low levels of nature connectedness.[14] Turning to more positive things, it has been found that a close relationship with nature is associated with more innovative and holistic thinking.[15]

Feeling good and functioning well

From the handful of nature connection studies at the turn of the millennium, there was a rapid growth in research. Soon, there were enough studies for 'systematic reviews' and 'meta-analyses'. These gather together the results from many research studies and provide a powerful summary of the evidence. In 2019 a PhD student I supervised published a review of nature connectedness and wellbeing in the *Journal of Happiness Studies*.[16] It pooled findings from 50 research studies involving over 16,000 people. Each study explored the links between nature connection and two broad types of wellbeing – feeling good and functioning well.

As you'd expect, feeling good includes positive emotions from pleasant experiences or fulfilling desires; the more technical term is 'hedonic wellbeing'. Hedonic wellbeing is more than feeling happy. There are many 'feel-good' emotions, ones that are high-arousal such as joy and those that are low-arousal such as calm. When we have a balance of both, we may experience higher levels of life satisfaction. Functioning well refers to those factors of wellbeing that help us see through each day, even when pleasant experiences are limited – things like the growth, self-acceptance and autonomy involved in leading a meaningful and purposeful life. The technical term is 'eudaimonic wellbeing'. It's also worth noting that the 'nature-inspired autonomy' linked to eudaimonic wellbeing suggests nature has value in allowing one to gain a personal sense of freedom and escape from the dissatisfaction that can come from the influences imposed by society.[17] As we've seen, societal influences and the pursuit of material happiness are likely involved in our broken relationship with nature.

Hedonic and eudaimonic wellbeing are related but distinct from each other. They are linked to different behaviours, motivations and experiences. When we have high levels of both, we are flourishing, and that's a great place to be. The meta-analysis of the 50 studies showed that people who are more connected to nature are more likely to be flourishing. Although it isn't possible to infer causation from the meta-analysis, we can do so from other research we'll consider in the next chapter. If we improve people's nature connectedness, we find their wellbeing improves too. In the meta-analysis, personal growth had a particularly strong relationship with nature connectedness. Nature can bring genuinely uplifting experiences, full of awe and wonder, which are associated with transcendence – events that move and change us. Finding awe and wonder in nature can expand our thoughts and foster personal growth.

Voices from the past

Somehow, we have forgotten this, but there is a long history of finding wonder in and being moved by nature. This history also includes a recognition that nature is good for us. In the opening chapter, I used some writing by Richard Jefferies to help envisage how alive and colourful the landscape once was. In the language of his time, Jefferies saw awe and wonder in nature and predicted that nature connectedness would bring happiness.

> Fanning so swiftly, the wasp's wings are but just visible as he passes; did he pause, the light would be apparent through their texture. On the wings of the dragon-fly as he hovers an instant before he darts there is a prismatic gleam. These wing textures are even more delicate than the minute filaments on a swallow's quill, more delicate than the pollen of a flower. They are formed of matter indeed, but how exquisitely it is resolved into the means and organs of life!

> From the tiny mottled egg come the wings that by-and-by shall pass the immense sea. It is in this marvellous transformation of clods and cold matter into living things that the joy and the hope of summer reside. Every blade of grass, each leaf, each separate floret and petal, is an inscription speaking of hope. Consider the grasses and the oaks, the swallows, the sweet blue butterfly – they are one and all a sign and token showing before our eyes earth made

> into life. So that my hope becomes as broad
> as the horizon afar, reiterated by every leaf,
> sung on every bough, reflected in the gleam
> of every flower. There is so much for us yet to
> come, so much to be gathered, and enjoyed.
> Not for you or me, now, but for our race, who
> will ultimately use this magical secret for their
> happiness. Earth holds secrets enough to give
> them the life of the fabled Immortals. My
> heart is fixed firm and stable in the belief that
> ultimately the sunshine and the summer, the
> flowers and the azure sky, shall become, as it
> were, interwoven into man's existence.[18]

As we've seen in Chapter 4 on the hidden connections with nature, the interconnectedness of humans and the rest of the natural world has become a secret – and there are magical examples to be found of how nature helps manage our moods to bring happiness. Later, in the final part of the book, we will explore how nature can be interwoven back into our existence.

Comparing exposure and connection to nature

Despite the historical awareness and intuitive sense, our fractured relationship is such that a great deal of money has been spent on research that shows exposure to nature is good for wellbeing. With growing evidence for the benefits of nature connectedness, there was a need to see what contribution connection makes. Was it as important as other benchmarks, such as socio-economic status? And was it exposure or connection that best explained wellbeing? As the benefits of nature become increasingly recognised, there is need for more work towards nature-based solutions for health. However, there is also a need to know if efforts are best directed at increasing visits and time in nature, or whether aiming to develop a closer relationship with nature could best maximise the benefits.

'Cross-sectional' research that explores relationships between various factors and how they contribute to and explain outcomes, such as wellbeing, is a handy way to explore such questions, but it is often dismissed for not showing 'causation'. There is an important distinction to understand when considering research evidence. Different types of approach are needed to answer different types of question. Different approaches have different purposes, so research that considers relationships and correlation does not provide evidence of causation. This research isn't used to establish causation; it is used to establish whether there are relationships and estimate contribution. If there isn't a relationship, it's probably not worth going to the trouble of looking for causation, as gathering evidence takes time, and that costs money.

The term 'cross-sectional' always makes me think of cake, and you might – well, I might – think of cross-sectional research as taking a slice through a mystery cake. By cutting a cross-section, we can see what ingredients contribute to the deliciousness, perhaps how many walnuts and raisins there are. We could then compare lots of different cakes, or their recipes if we don't want to get into trouble at the supermarket, and see how the various levels of dried fruit and nuts relate to tastiness. If we found walnuts do not relate to tastiness, then there's no need to establish causality. However, if we found a relationship between raisins and tastiness, then to establish a causal relationship between raisins and tastiness, we would need to approach things differently and use an experiment to control and compare cakes with and without raisins. Or we might find that raisin levels seem to explain twice as much tastiness as walnuts and recommend, once causation is established, adding raisins to more cake recipes. The same applies to recipes for wellbeing; by comparing time in nature and nature connectedness, we can make recommendations on the best approach to take, referring to empirical research that has established causality along the way.

Now that I've neatly recovered that tangent from a potential cul-de-sac, let's return to the contribution of nature connectedness to eudaimonic wellbeing, specifically the feeling of living a meaningful and worthwhile life. Meaning comes from connection, yet the Western dualistic tradition has separated humans from that feeling of oneness with the wider natural world. This leaves many with a sense of meaninglessness where no level of power, wealth, possessions or fame will bring satisfaction – which fuels unsustainable consumption of natural resources and perpetual growth that divides the haves and the have-nots.[19] So, establishing the factors contributing to a meaningful life could also bring about more sustainable and equitable living. Perhaps nature connectedness and a new relationship with nature is the missing link in a sustainable and worthwhile life?

Beyond nature contact to connection

Earlier I introduced the Nature Connection Index, designed as a population measure of nature connectedness. The project was coordinated by Natural England, supported by several national nature conservation organisations, and involved a number of universities along the way. Data was collected from a representative sample of 4,960 adults across England. The project aimed to establish the contribution of nature connection and nature contact to wellbeing.

This cross-sectional research investigated the relationships between nature connectedness, general health, wellbeing and nature contact, in the form of visits and neighbourhood green spaces, within a single study. In the analysis, steps were taken to account for a comprehensive range of social and demographic variables – factors such as gender, ethnicity, employment, marital status, socio-economic status, neighbourhood deprivation and urbanicity. This complex analysis found a positive association between nature connectedness and feeling that one's life has meaning and is worthwhile. This contribution was nearly four times larger than the increase associated with higher socio-economic status, which of course is strongly linked to wealth. In other words, a close relationship with nature was four times more important than purchasing power when it came to feeling life has meaning and is worthwhile.

Furthermore, nature connectedness was important over and above nature visits, the measure of simple contact with nature. However, the level of nature visits was more important for physical rather than mental health, which makes sense. Interestingly, living in a greener neighbourhood was unrelated to wellbeing, highlighting the difference between the presence and use of green spaces.[20]

Of course, in science, we attempt to isolate and control variables, but in reality, everything is related, so the analysis also considered how the main factors worked together. Once again, nature connectedness was a key factor in interaction effects with nature contact. These interaction effects show that an individual's level of nature connectedness influences how they respond to contact with nature. For instance, frequency of visits to nature and nature connectedness interacted, suggesting optimal visits are likely to be those that also activate nature connectedness.

It is important to reiterate that the link between living a more worthwhile life and nature connectedness remained after accounting for a comprehensive range of social and demographic factors, and various types of nature exposure. This work showed that the contribution of nature connectedness is practically meaningful and greater in magnitude than benchmark socio-demographic factors such as socio-economic and marital status. Previously, research in nature and wellbeing has largely overlooked the person-specific factors in human–nature interactions. This work was the first to provide evidence that for eudaimonic wellbeing, nature connectedness is important over and above simply getting out into nature.

Moments not minutes

Although this study was nationally representative and well controlled, when establishing the benefits in a relatively new area, replication of research is welcome. This is especially the case when comparing to established benefits, such as looking at contact versus connection. Nature connectedness research has emerged more recently than nature exposure work. So, there are many possible factors to consider. Working with the National Trust, we explored how engaging with nature in simple ways and connecting to nature both relate to wellbeing.[21]

Once again, a nationally representative sample of just over 2,000 adults was surveyed to explore how various factors related to eudaimonic wellbeing (i.e. worthwhile life), hedonic wellbeing (i.e. happiness), ill-being (i.e. depression and anxiety) and general physical health. The factors that we considered were: time in nature, engagement with nature through simple everyday activities, knowledge and study of nature, indirect engagement with nature (e.g. watching nature documentaries and reading nature books), and, of course, nature connectedness. So, as explained, the key focus of this cross-sectional approach was examining these types of nature engagement simultaneously to reveal their relative importance, like comparing the contribution of walnuts and raisins to cake tastiness.

As ever, there was some complex statistical analysis, performed by the wonderful Canadian researcher Holli-Anne Passmore who joined us in the UK, bringing brightness to the bleak winter of 2019/20. A consistent pattern of results emerged across multiple analyses. We found that time in nature did not play a significant role in predicting wellbeing. Nor did indirect contact with nature, or knowledge and study of nature. Instead, both engaging with nature through simple activities, like smelling wildflowers, and nature connectedness consistently emerged as the prominent and significant factors in explaining mental health and wellbeing. When measured in concert, natural wellbeing is about moments, not minutes.

In that analysis, two essential components of a good life were also controlled for: the basic psychological needs of relationship status and autonomy (or having a sense of control over one's life). Even then, engaging with nature through simple activities and nature connectedness still had significant relationships with the feeling that life is worthwhile, with happiness and with lower rates of health issues such as depression and anxiety.

Wellbeing involves many complex factors, and this work helped confirm suggestions that nature connectedness and the lived experience of that connection, tuning into nature, are fundamental components of wellbeing and form a core psychological need. Given the long pre-history of the human–nature relationship and our roots in the forest, it would be more of a surprise if a close relationship with nature didn't matter for our wellbeing. Yet it's worth repeating; people accept that fish need the river, birds the sky and apes the forest, but question whether humans need nature. Such is our disconnection from the natural world.

Given the robust research that has demonstrated repeatedly that time in nature is important for our wellbeing, it may seem surprising that spending time in nature emerged as a lesser factor. However, previous research has tended not to include individual factors of nature engagement and connectedness. When these person-based factors are considered together, they have stronger relationships with wellbeing than time in nature. Previous research using time alone is likely to be tapping into various forms of nature engagement and connectedness by proxy, but not as well as using specific measures of engagement and connection. This is important, as the factors found to matter in research ultimately shape advice and even policy. Recommendations are based on what is measured.

To look at it another way, imagine that dietary research had focused on visits to the fridge and time spent eating. Dietary advice would focus on reducing those. That would be of some use – but it's what you eat, indeed your relationship with food, that matters. Including food-based factors such as salt, calories and fat content would mean time eating would become much less relevant. As noted earlier, there'll be strong relationships between time spent in hospital and recovery from illness, but we know it's the treatment received during that time that matters.

So, when thinking about wellbeing and nature, rather than focusing on visits and time in nature, consider what happens during those visits. Clearly, engaging with the natural world and having a close relationship with it generally involves spending time in nature. Yet time in nature may not involve active engagement with it. What matters is how that time is spent. Time does not tell the whole story of developing and being in a close relationship with nature – just as time spent together does not provide the full story of the closeness of human relationships.

Returning to the research, and getting technical for a moment, the survey data underwent 'dominance analysis' to look further into the role of time and connection. As the term suggests, dominance analysis considers if one factor dominates another in a relationship. It rarely reveals 'complete dominance',[22] yet here it showed that all other nature-related factors, including time and knowledge, were completely dominated by nature connectedness in explaining a life perceived as worthwhile and lower levels of ill-being. Engaging in simple nature activities ranked second. Meanwhile, engaging with nature through simple activities completely dominated all other nature-related factors when explaining happiness – with nature connectedness ranked second. This shows the powerful impact simple engagement and close connection with nature have on our wellbeing. Fish, birds and apes probably feel the same.

As we've seen, although science seeks to control and isolate variables, the reality of life is based on relationships. For this reason, the final approach in studying the survey data was 'commonality analysis'. Commonality analysis provides a fuller account by showing where factors work alone or together. Once again, engaging in simple nature activities and nature connectedness emerged as foremost in explaining levels of wellbeing. Both uniquely explained happiness, a worthwhile life and lower ill-being far more than time in nature. Indeed, only 1% of the variation in these measures was uniquely explained by time in nature. Time in nature is of course not wasted, however – as well as being essential for engaging with nature to build connection, it is time in nature that best indicates levels of physical activity and, therefore, physical health. Hence, when turning to general physical health, time in nature emerged as significant, but wait for it – alongside nature connectedness. When moving to the dominance analysis again for physical health, there was a tie between nature connectedness and time in nature.

Here we can see that a consistent pattern emerged across all forms of analyses. This strongly suggests that a close relationship with nature is important for our health and wellbeing, and that time in nature is not the main factor in the nature–wellbeing link. If we are not tuned into nature each day, our lives are poorer in terms of meaning and happiness. And tuning into nature is not about minutes. It's about moments. Moments engaging with nature in simple ways and developing a close connection with nature explain wellbeing better than time spent in nature.

This finding that nature connection explains wellbeing better than time spent in nature was then supported in another study by a group of Chinese researchers working with data from a Chinese population. Similarly, they found that nature connectedness was a stronger predictor of mental wellbeing outcomes than nature exposure. Nature visits weren't linked with wellbeing. The only exception was for ill-being or depression, where visit frequency made a greater contribution, although nature connectedness was also involved in this relationship. People with higher levels of nature connectedness benefit more from nature contact. The researchers concluded that nature visits can help restore people from ill-health, but their work did not find a contribution to promoting mental wellbeing. The researchers also recommended a shift in policy-making to target nature connectedness, in addition to providing a greener environment.[23]

Lessons from lockdown

Unexpectedly, the 2020 restrictions used in an attempt to control the coronavirus pandemic brought us those all-important moments. Later in the book, when we move on to our work to develop interventions to improve nature connectedness, we'll see that such is our disconnection from nature that we had to pay people to take a moment with it! The lockdown restrictions forced many people to slow down, and being outside was one of the few things to do. This provided an opportunity to explore how nature, and a close relationship with it, benefited the nation's mental wellbeing. Handily, March 2020 saw the launch of the new People and Nature Survey developed by Natural England. This provided valuable data, especially as the new survey included a measure of nature connectedness and noticing nature.[24]

Building on the two studies above, we compared how people's pre-pandemic, longer-term nature-visiting habits, relation-ship with nature, recent visits and noticing of nature explained wellbeing. We also included how the restrictions contributed to nature's wellbeing through people's pro-nature behaviours, but we'll consider those results later. The People and Nature Survey got responses on these aspects from several thousand people. It showed that nature visits increased by 40% during the restrictions – a fact that has been often reported. Less commonly reported is that noticing nature increased by 74%. The survey also collected information on happiness, life satisfaction and the sense that one's life is worthwhile.

When considering people's longer-term habits and lockdown behaviour together, the analysis showed that increases in nature visits during the restrictions didn't account for higher levels of wellbeing over and above that linked to visiting habits before the pandemic. However, higher levels of noticing nature during the restrictions were associated with a greater sense of having a worthwhile life. We then turned our attention to people's engagement with nature during the pandemic restrictions. We found that both increases in noticing nature and the number of recent visits to green spaces explained higher levels of happiness, life satisfaction and the sense of life as worthwhile – with noticing nature tending to contribute a little more than visits.

Similarly to the work above, these findings highlighted the importance of having a long-term relationship with nature and actively noticing it. This research, supported by the empirical work we'll consider later, suggests that noticing nature can bring wellbeing benefits regardless of one's historical relationship with nature.

Sadly, although many people found a friend in nature during the 2020 restrictions, data showed that this was just a short-term relationship for some. As more normal living returned, that relationship was stolen again, with the levels of nature connectedness falling by 25% between April 2020 and April 2021. Interestingly, it wasn't that people had stopped visiting nature. The number of people visiting green and natural spaces stayed higher than it had been, but once people were free to engage in other activities, the rise in 'noticing nature' fell away. This also reminds us that visiting nature does not necessarily bring connection. For many, nature is only a background for other activities.

During this chapter, we've seen that a clear picture is emerging from three large-scale national datasets. A close relationship with nature and actively engaging with nature explain our levels of wellbeing. Time and visits to nature have a role – to notice, we need to visit – but to maximise the benefits, they should not be the focus. A close relationship and simple engagement with nature bring benefits over and above those derived from simply spending time in nature. Focusing on time and visits to nature is a more straightforward box to tick; a new park can provide a local green space or outreach can help more people visit the natural landscape. Yet, as we'll see, building a close relationship with nature requires a different approach. Relationships are complex, and a focus on encouraging and prescribing visits can reduce the wellbeing benefits and increase levels of anxiety, with pressure to spend time in nature reducing motivation levels.[25]

The focus on moments over minutes is not about undermining research based on measuring time and visits. That work has been essential and valuable in establishing the benefits of nature for wellbeing – in showing us that humans are like fish, birds and apes. However, it is crucial to think about relationships for a sustainable future that unites human and nature's wellbeing. We've seen the power of metaphors and language; referring to 'spending time in nature' can seem transactional. A 'dose of nature' suggests nature as a pill to pop when things are not so good, straight from the dualistic biomedical model that separates us from the spoonful of nature that returns us to 'normal'. A dose of nature also suggests continuing a culture of occasional visits to 'special' green spaces, but nature should not have a 'part-time role' in wellbeing. If we do need a dose of nature, we need it like we need a dose of air: always. Each day, nature should have a full-time role through ordinary engagement, through creating moments to notice nature on the doorstep. Such a world would also deliver the health benefits of living close to nature, but furthermore, fostering reconnection with the rest of the natural world is essential for a sustainable future.

CHAPTER 7

How Does Reconnection Bring Wellbeing?

Satisfying our innate union

Studies showing that noticing nature is linked to improved wellbeing, and talk of a dose of nature for restoration, can turn our thinking to the mechanisms by which a close connection with nature brings wellbeing. Having an idea of *how* this close relationship improves wellbeing helps develop effective interventions such as green prescriptions, and promote the wellbeing benefits of nature connectedness. More widely, in the context of the crisis of the loss of wildlife, it is important to explain how a healthy natural world matters for human wellbeing.

There are various theories that attempt to explain the wellbeing benefits of nature and a connection to it. Connection goes beyond the restoration provided by visiting nature, fulfilling deeper and essential needs that develop and maintain emotional wellness. Earlier, we saw how being in, viewing and touching nature could manage our emotions. It has been argued that a close relationship with nature satisfies the innate union humans have with the rest of the natural world, that bond denied by modern urban living.[1] It is thought that being close to nature and the wider events of the natural world provides perspective on personal feelings, improving acceptance, emotional development and self-awareness.[2] Another theory maintains that the experience of being part of nature is essential, providing an opportunity to show care, which is a core to emotional wellness.[3] Indeed, it is suggested that experiencing the relationship of being part of nature is the only way to achieve optimal wellness.[4] I feel this in simple ways such as being with the community of birds in the garden. Witnessing and caring about the lives led around me provides wider context for personal feelings.

This returns us to birds, fish and apes; for them, optimal wellness is in nature because it is the habitat they evolved in. Birds and fish are some evolutionary distance away. However, the brain structures linked to anxiety disorders are shared by a range of animals, and there is a psychological, social and biological continuity between the great apes[5] – to the extent that there are similarities in vulnerability to stress and mental health issues between humans and great apes.[6] Take apes away from their natural habitat, and abnormal behaviours and disorders associated with humans become apparent. Chimpanzees are our closest living relatives, and a study of 40 socially housed chimpanzees living in zoos analysed 1,200 hours of behavioural data. It concluded that abnormal behaviour was endemic despite efforts to enrich their artificial homes.[7] Compared to those in rich natural habitats, chimpanzees living in zoos occupy more limited environments, and large parts of their lives are managed. It has also been found that when poorly housed chimpanzees return to more naturalistic enclosures, abnormal behaviours decrease, and normal behaviour patterns return.[8] Poorly treated chimpanzees can have mental health issues we're familiar with, such as obsessive-compulsive and anxiety disorders, which aren't typically observed in wild chimpanzees.[9] Indeed, a study of 196 wild chimpanzees estimated rates of depression in chimpanzees at 3% compared to 17% in UK adults, and PTSD at 0.5% in chimpanzees compared to just over 4% in people.[10] Clearly, such estimates are difficult and captivity itself is a major factor. Still, it is interesting to note how mental health disorders common in humans aren't typical in the natural habitat, but are observed in more artificial environments.

Wellbeing through beauty and emotions

There has been little specific research into how nature connectedness leads to improved mental wellbeing, but there are theories on the benefits of exposure to nature. Indeed, there is evidence that nature connectedness is a mechanism through which nature exposure brings wellbeing.[11] However, the main models used to explain these wellbeing benefits are based upon psychological restoration.[12] Attention Restoration Theory and Stress Recovery Theory provide important accounts of how nature restores us when we're tired or stressed.[13] However, early indications have suggested that these two theories may not explain the benefits of nature connectedness.[14] Furthermore, it's not all about restoration; nature is also good for us when we're not feeling run-down.[15] People seek out nature to manage their moods when happy and sad.[16]

We've seen how simply viewing beautiful flowers brings beneficial physiological and emotional changes to humans. Research has considered the link between nature's beauty, nature connectedness and wellbeing. In two studies, the positive relationship between nature connection and wellbeing was only significant for those who tuned into and engaged with nature's beauty.[17] However, others looking at this relationship found these effects weren't universal – they were present in Canadian samples, but not in Japanese and Russian data. That said, this research identified evidence for a subtly different route. That, rather than connection bringing wellbeing via engaging with nature's beauty, the wellbeing benefits are explained by engaging with nature's beauty, leading to enhanced nature connectedness.[18] For everyday understanding, these nuances don't matter a great deal, engaging with nature's beauty is likely a good thing to do!

To look into how nature connectedness might bring about wellbeing, we explored the role of engaging with nature's beauty and emotion regulation in a couple of studies.[19] Firstly, using data from the Wildlife Trusts' successful '30 Days Wild' campaign, we asked how strongly people responded to nature's beauty to explore the role this played in the link between nature connectedness and wellbeing. In a second study, we measured how well people regulated their emotions – for example, how easy or difficult they found it to regain control after becoming upset. Once again, this was to explore the role that emotional regulation played in the link between nature connectedness and wellbeing.

A close look at the numbers revealed that taking part in 30 Days Wild improved nature connectedness, wellbeing and engagement with the beauty of nature. The increases in nature connectedness brought wellbeing improvements through that increased engagement with nature's beauty. Similarly, in the second study better emotional regulation played a part in the link between nature connectedness and wellbeing. People who found emotional regulation more difficult and struggled to manage their moods had a lower connection with nature, and lower happiness. This was the first evidence linking emotion regulation to the wellbeing benefits of nature connectedness. Since then, others have considered this relationship – finding, for instance, that a close relationship with nature was linked to lower stress levels through strategies to manage emotions via acceptance of situations.[20]

Taken together, these studies suggest that both engaging with nature's beauty and emotional regulation play a part in the wellbeing benefits nature connectedness brings. Interestingly, although both have a role in the relationship between nature connectedness and wellbeing, difficulty in emotional regulation was not associated with difficulty in engaging with nature's beauty. This suggests that a close relationship with nature brings wellbeing in two different ways, potentially through cognitive mechanisms such as processing fluency and being attuned to nature's beauty[21] – as well as through affect regulation. As mentioned earlier, the three-circle model suggests that engaging closely with nature can calm our emotions, bringing the emotional balance needed for wellbeing.

Meeting spiritual needs

However, the ways in which nature connectedness brings wellbeing are likely to be complex and involve multiple factors. For example, several researchers have suggested that Self Determination Theory (SDT) provides potential mechanisms.[22] SDT covers people's basic psychological needs, with autonomy, competence and relatedness being considered essential for wellbeing. A close relationship with nature can satisfy some of these needs, such as relatedness.

Another potential way that nature connectedness brings wellbeing is via spirituality. Spirituality is a big notion that can be easier to shy away from. To avoid getting bogged down, let's go with spirituality relating to our inner experience and beliefs that give meaning to existence beyond the here and now. Earlier, we touched on Indigenous peoples' strong spiritual connection to nature, and when the role of spirituality has been considered, it has been found to play an important role in the link between connection to nature and wellbeing – and this was in university students.[23] Along similar lines, other work has shown that finding meaning in life is a likely part of the positive impact on wellbeing. Improving our relationship with nature brings meaning to life, and knowing that our lives are meaningful and transcend the present matters for our wellbeing.[24] It is thought that a close relationship with nature plays a fundamental role in tackling several existential anxieties[25] – such as those around big topics such as our identity, freedom, happiness and death. Fully flourishing and purposeful living requires a close relationship with nature to balance our desire to focus on income and material gains as the primary source of success and wellbeing.

The role of this sense of purpose and worthwhile existence in the link between nature connectedness and wellbeing was explored further in an interesting study in the Philippines.[26] A great deal of nature connectedness research, and psychological research in general, takes place in Western populations, so this work also provides evidence of the universality of the human need for nature and the impact of increased urbanisation. As well as featuring a non-Western population, it is also interesting that this work took place in the context of ongoing environmental degradation. The Philippines has seen a rapid loss of forests, and deforestation continues with ongoing extinction of nature experience through high rates of migration to urban areas.[27] Further, the Philippines is a leading source of plastic waste in the oceans.[28] The research found that people with a closer relationship to nature sensed greater purpose and meaning in their lives, leading to greater mental wellbeing. The study also found that when people placed greater importance on material gains, they lacked the presence of meaning in their lives, which was linked to diminished wellbeing. So, returning to mechanisms, the work showed the role of meaning in the link between nature connectedness and wellbeing.

This chapter has considered some quite theoretical research. However, at heart the story is simple. As far as we know, chimpanzees aren't consciously engaging with beauty, meeting spiritual needs and seeking meaningful lives. However, they are embedded within nature's beauty with the competence to live autonomous lives. Take that connection and meaning away and emotions can become unbalanced, and wellbeing suffers. For humans, meaning can then turn towards material gains in an urban environment. The research shows the need, in living this life, to engage in an affective relationship – to notice, become sensitive to nature's beauty and find deeper meaning to access the wider benefits of nature connectedness and wellbeing. We'll return to the practical steps to improving the human–nature relationship later. A fundamental point missing from our present discussion, however, is that there can be no human wellbeing without nature's wellbeing. Hence, we'll now move on to consider the benefits of nature connectedness for the natural world.

CHAPTER 8

Good for Nature: Environmental Benefits of Reconnection

The link to pro-environmental actions

The warming climate and loss of wildlife are caused by human actions, resulting from a broken relationship between people and the rest of nature. Together, through this book, we've wandered through the darker territory of the reasons for that broken relationship and seen the hope of our innate connections with the wider natural world that can happen on our doorstep. Then we were introduced to the science of nature connectedness and the benefits a close relationship with nature brings for human wellbeing. Now we'll consider the benefits that a closer relationship with nature brings for nature itself. Research shows that increased levels of nature connectedness lead to more pro-environmental behaviours, often referred to in the trade as 'PEBs'. That's why to help address climate change and nature's recovery, we need a closer relationship with nature.

Earlier, we considered how systematic reviews that bring together many independent research studies provide robust evidence of the link between nature connectedness and human wellbeing. Around the same time, a systematic review and meta-analysis showed the link between nature connectedness and pro-environmental behaviours. Caroline Mackay and Michael Schmitt from Simon Fraser University in Canada brought together 92 separate research studies involving over 27,000 people.[1] They found 'compelling evidence for a strong and robust association between nature connection and PEB'. Importantly, they also found a significant causal effect, with a closer relationship with nature leading to pro-environmental behaviours. They concluded that targeting nature connectedness is a promising avenue for promoting these behaviours. At around the same time, researchers in New Zealand were conducting their own systematic review and meta-analysis. They found a similar relationship between connection to nature and pro-environmental behaviours.[2] They also noted that this relationship was ubiquitous across countries and cultures. So, there is good evidence that improving nature connectedness is a route to both human and nature's wellbeing. With the United Nations and IPBES recognising that the climate and biodiversity emergencies require a new relationship with nature, the science of nature connectedness offers an evidence-based direction to take.[3]

Much of the nature connection research has focused on human wellbeing, but we have also considered the benefits of a closer connection with nature for nature's wellbeing. Just as human actions cause harm to nature, benefits can come from many different actions too. There are many types of pro-environmental behaviour; there are many environmental problems! We did some work looking at how different behaviours were linked to nature connectedness. This also provided an initial suggestion of how much closer the human–nature relationship needs to be. Earlier, we saw how the Nature Connection Index was included in a national survey of a few thousand people, which gave an average level of nature connectedness of 61 across the population. We saw how this varied a little between various communities, and a lot over a person's lifespan with the 'teenage dip'. That work also considered the pro-environmental behaviours required for a healthy natural world.[4] We found a near-perfect correlation between nature connectedness levels and the percentage of people participating in the environmental behaviours we considered. This strongly suggests that the behaviours requiring greater commitment, indicated by fewer people doing them, are linked to a closer relationship with nature.

Some examples help illustrate this link. We found that for those who engage in the relatively straightforward behaviour of recycling, the average level of nature connectedness was 63.

Recycling is made easy for many with specific bins and regular collections from home. However, giving up time to volunteer to help the environment takes much more commitment, and for the 5% of people who reported doing this, the average nature connectedness score was 76. There was also an average of 76 for those most concerned about damage to the natural environment. It was clear that those people involved in pro-environmental behaviours have a closer connection with nature, and we tentatively set a nature connectedness target for a sustainable future of over 70. That's at least 15% above the current population average in the UK – with a 25% increase needed to reach 76, the level associated with the meaningful behaviours and attitudes. Such a change would make a sustainable future more acceptable to people and more likely. There has been a more detailed analysis of such thresholds of nature connectedness, but with children, where it was found that current average levels were again close to 15% below the higher levels that might be needed.[5] We'll look at the ways to improve nature connectedness later, but the good news is it can be done, and sustained increases of that magnitude have been achieved.

From carbon cutting to habitat creating

So far, we've been referring to pro-environmental behaviours. Yet there are two environmental crises: climate change and biodiversity loss. There is a relationship between them, but they are different; the UN has separate conventions on biological diversity and climate change. Sadly, as we touched on earlier, there is a considerable disparity in awareness, coverage and research. For example, media coverage is around eight times higher for climate change, and there's less funding and research into biodiversity loss.[6] This is also reflected in psychology research, as nature conservation behaviours directly relevant to biodiversity, such as planting pollinator-friendly plants, are often absent from the pro-environmental behaviour measurement scales. These scales tend to focus on actions that can reduce carbon and resource use, such as recycling or using public transport. The bias towards climate change is such that some psychologists would argue that pro-environmental behaviour measures are sufficient and pro-nature conservation behaviour measures aren't needed. But in the real world, nature conservation organisations see that pro-nature conservation behaviours are different.[7] Plus, when I've worked with the RSPB and the Wildlife Trusts, they were keen to measure pro-nature conservation behaviours specifically.

Our work using the population measure of nature connectedness allowed us to look at this difference between nature conservation and environmental behaviours in more detail.[8] This confirmed an important point. In addition to having different purposes, pro-environmental and pro-nature conservation behaviours were distinct types of human behaviour that should be thought of differently. So, pro-environmental behaviours broadly related to carbon and resource use are distinct from pro-nature conservation behaviours that are broadly related to habitat creation. This was good news, as we had previously spotted the issue and already started developing the first validated pro-nature conservation behaviour scale – more on those results later.

Firstly, we should consider more of the results from the population survey work. We found that people were doing far more pro-environmental behaviours, such as recycling, than pro-nature conservation behaviours. We've already seen how different behaviours require greater commitment and are associated with higher levels of nature connectedness. Just as we've seen for human wellbeing, this work also allowed us to compare how contact and connection with nature explain both pro-environmental behaviours and pro-nature conservation behaviours.

We found that people visiting nature once per week or more tended to enact more pro-environmental behaviours such as recycling, buying eco-friendly products, walking or cycling. However, we didn't find a link between visits and pro-nature conservation behaviours, such as supporting nature conservation organisations and volunteering. We also found that living in a greener neighbourhood was unrelated to either of the pro-nature behaviours. In contrast, people who had a closer relationship with nature took part in more pro-environmental and pro-nature conservation behaviours. When we compared the two, a person's relationship with nature explained levels of pro-nature conservation behaviours and pro-environmental behaviours better than people's frequency of nature visits. So it's right to focus on the human–nature relationship when considering climate warming and biodiversity loss.

Pro-nature conservation behaviours

Let's look more closely at pro-nature conservation behaviours, as they should be a core concern for those with a close relationship with nature. Indeed, they should be a core concern for all, given that ultimately there can be no human wellbeing without nature's wellbeing. The focus on climate change over wildlife loss is odd. As we considered earlier, we don't yet tend to have a relationship with the climate and climate change; although currently very real for some, it is an intangible threat decades away for many. The loss of most of wildlife since 1970, on the other hand, is tangible and real. The actions people can take to make a difference in terms of wildlife loss are also more positive than the positive inaction required to help make a difference to climate change. For example, creating homes for nature locally is a positive action that brings the rewards of local wildlife. Whereas using less hot water is positive inaction – doing less for no visible reward.

Pro-nature conservation behaviours are also now of more interest to nature conservation organisations. Clearly, nature conservation action is a fundamental purpose of these organisations, but approaches have changed. There has been a move from protecting sometimes isolated nature reserves towards nature recovery networks.[9] In urban and residential areas, private gardens often provide the largest area of green spaces, and widespread individual action can have value, helping to overcome the fragmentation of habitats.[10] Even at a landscape scale, the behaviour of individuals is still essential.[11] One landowner can take action at scale, from full-scale rewilding to more nature-friendly land management. Here the relevance of factors at the heart of nature connectedness has been identified as a new direction for nature conservation.[12]

Finding effective ways to address the loss of wildlife and increasing pro-nature conservation behaviours requires better understanding and measurement. This requires a reliable and valid measurement tool, so as noted earlier, we set out to develop one – namely, the Pro-Nature Conservation Behaviour Scale, or ProCoBS for short. Developing a psychometrically validated scale goes far beyond simply thinking up some questions and adding up the responses. We began by creating an 'item pool' – a long list of all sorts of pro-nature conservation actions people can do, either at home, as civil actions, or in the land or gardens people look after. This long list of 42 actions was refined by asking 70 ecologists which were most important. The most important actions people could do at home included providing food, water and nesting boxes for birds and bats. In comparison, returning insects found in the home to the outdoors was not seen as important. The ratings from the 70 ecologists suggested that important civil actions centred around volunteering for conservation organisations and supporting nature-friendly legislation. Activism, demonstrations and sharing conservation articles were seen as less important. Finally, for garden and land management, planting pollinator-friendly plants and leaving undisturbed areas scored highly, whereas avoiding planting exotic plants did not.[13] Interestingly, nature conservation journals had no interest in publishing this expert-ranked menu of nature conservation actions, despite the lack of information often cited by people as a barrier to helping nature.[14]

The most meaningful actions as rated by the ecologists were then used to develop the nature conservation behaviour scale. Three hundred people completed this early version, and it underwent some clever psychometric scale development analyses to select the best indicators to include in the scale. Then 250 of them completed it again to establish that the measure was reliable and gave the same results. In a third study, thanks to interest in the measure from the National Trust, a representative sample of over 1,000 adults in the UK completed the scale. This allowed further confirmatory analysis to show it was all working nicely, or more formally that it was a psychometrically robust measure.[15] Having a working scale allows some fundamentals to be established. For instance, women and girls showed higher engagement in pro-nature conservation actions than men. Looking at ages, there was a dip in the teenage years, similar to that found for nature connectedness. Clearly, teenagers tend not to own their own gardens, but this dip was also found for civil actions. As might be expected, people living in more rural areas engaged in more pro-nature actions.

From start to journal submission, developing ProCoBS took 18 months of a PhD, which was relatively quick. Only for the scale to be robustly challenged as unnecessary by one reviewer; such is the researcher's life! Although some psychologists think that pro-nature conservation behaviours are no different to pro-environmental behaviours, the proof comes in the interest in the pudding. ProCoBS is being used by nature conservation organisations to inform their reports and campaigns, and nationally in Natural England's People and Nature Survey.

There is good reason to focus on pro-nature behaviours. Although the stark decline in wildlife populations paints a very depressing picture, the fact that they are linked to human actions means we can act to reverse current trends. People can make a difference in their gardens or local green spaces by supporting wildlife. Simple changes can improve habitats and increase biodiversity, making gardens an important tool for nature conservation. There can be resident wildlife management of local parks for those without a garden, and not all pro-nature behaviours involve actively improving habitats. People can be political, influencing public policy decision-making and social change. Having a tool for measuring people's nature conservation behaviours means initiatives and progress can be tracked.

Measurement tools are rather technical. It's worth closing this chapter by reminding ourselves why nature conservation behaviours matter. In *Framing Nature*, conservationist Laurence Rose explores the human relationship with the rest of the natural world.[16] He reminds us of the stories of individual species and their interconnectedness with human activity. How the corncrake benefited from settled agriculture, becoming common, before becoming the first victim of the changes in and then industrialisation of agriculture. A loss accepted and now fading from memory, to be replaced by new losses that are accepted once again, such as the lapwing. This is a bird I recall flocking on the school playing field as a child; now I watch a lone pair describe their calls through the patterns of their flight. Loss is normal, but this is not loss; they are not missing or misplaced. They have been decimated and taken from us, by us. Rose reminds us that these absent birds were once part of the everyday experience that shaped our lives. When we notice and get to know wildlife, they become fellow beings, friends that visit us. Wildlife that could either be the next to be taken and forgotten, or the spark of a reconnection with nature.

CHAPTER 9

One Health

What explains care for nature?

Nature's recovery depends upon more people taking more action to reverse the loss of wildlife. To understand what motivates people to do more for nature, measures of pro-nature actions are needed. The evidence from these measures allows effective interventions and communication programmes to be developed and evaluated. It is crucial we know what works best.

The first research into the factors that explain the pro-nature actions of nature lovers involved a representative survey of 1,298 adults in the UK.[1] We looked at many potential factors: time spent in nature, active engagement with nature, indirect engagement with nature through reading books and watching documentaries, knowledge and study of nature, valuing nature, concern for nature, pro-environmental behaviours and, you must have guessed, nature connectedness. Through some neat analysis, we found that altogether these factors explained 70% of the variation in people's actions for nature. That is an unusually high figure. The factors that best explained people's pro-nature conservation behaviours centred on their connection with nature: levels of nature connectedness and active engagement with nature through simple things like listening to birdsong. Watching nature television programmes and reading nature books was also a significant factor, along with engaging with broadly carbon-cutting pro-environmental behaviours.

We then compared the relative importance of all these factors in explaining pro-nature actions. The amount of time people reported spending in nature and knowledge about or study of nature did not emerge as significant – there wasn't a relationship.

Nor, surprisingly, did valuing and having concern for nature; we'll return to that. Instead, the lived experience of nature connectedness, engaging in simple nature activities, emerged as the most important contributor to pro-nature conservation behaviour. However, as we've noted, the real world is all about relationships; the factors we try to isolate for research rarely work alone. So, we again used commonality analysis to reveal that just 15% of the variance in people's pro-nature behaviours was purely explained by factors working alone. However, a person's relationship with nature, their nature connectedness and their engagement in simple activities played a part 92% of the time.

The message was clear: our efforts to encourage people to do more for nature should focus on nature connectedness and simple everyday engagement with nature, the nature that is disappearing. As we saw earlier, for human wellbeing, rather than being a simple matter of time, visits to natural spaces must involve simple engagement with nature and activities that increase nature connectedness. Studying and sharing knowledge about nature is not enough – another topic we'll return to. Speaking after the publication of the IPBES landmark health-check of life on Earth, Sir Bob Watson, lead scientist of the work, noted that the core issue in biodiversity loss was humans. Therefore, we need to ask how we can become more in tune with nature. How do we relate to nature?[2] The science of nature connectedness can help answer these questions.

As we've seen, the biodiversity crisis shows that the human relationship with nature is broken. The above-mentioned research shows the steps required to help get more people doing more for nature. It was the first of its type and only a single study. Still, together with the wider evidence, it provides important direction for local initiatives and broader policies, especially as the time for nature's recovery is tight.

The importance of nature connectedness for pro-nature behaviours is also emerging in a range of other research. For example, people with a closer connection with nature are more willing to pay for conservation work in the forests of Singapore.[3] Focusing on pollinator conservation behaviour, Jessica Knapp and colleagues found a close relationship with nature was an important predictor, alongside feeling able to help and, to a more limited extent, knowledge.[4] However, the relative importance of each varied between conservation actions. Similarly to the importance of simple engagement with nature, they found the diversity of nature interactions was also important. As we know, an emotional relationship with nature is a key part of nature connectedness, and a joyful response to pollinators has been found to predict pro-pollinator behaviours.[5]

As we saw earlier, the 2020 coronavirus restrictions increased how often people noticed nature and increased nature connectedness. As well as looking at how this reconnection with nature explained the nation's mental wellbeing, we considered how it linked through to pro-nature conservation behaviours.[6] Once again, we were enabled by the new People and Nature Survey developed by Natural England, which included our ProCoBS scale. Looking at the longer-term pre-pandemic relationships, we found that people who spent more time in nature and had a greater connection to nature reported more pro-nature conservation behaviours. Then, turning to the short term, increases in noticing nature during the restrictions, but not recent visits, explained higher levels of pro-nature conservation behaviours. In sum, higher levels of pro-nature behaviours were best explained by higher levels of noticing nature and nature connectedness.

After confirming the importance of noticing nature, we wondered if there was a feedback loop – that is, if taking steps to conserve biodiversity might also connect people to nature more strongly.[7] For there to be such positive feedback, people need to notice the results of their actions through increasing visible biodiversity. This increased sensory contact with the natural world might lead to a closer relationship with nature. The various pro-nature actions we can take vary in how much visual feedback they provide. For example, a sunny patch of pollinator-friendly flowers will likely provide more visible biodiversity than a shady log pile under a hedge. Once more, we turned to the People and Nature Survey in the UK. We looked at responses from over 4,000 people, and as expected, the more visible conservation actions that provided more feedback predicted how much the participants reported noticing nature. In contrast, the conservation actions that produced less visible biodiversity and feedback were unrelated to noticing nature. To complete the loop, we found that the levels of noticing nature people reported were also positively related to their levels of nature connectedness. Plus, pro-nature actions linked to people noticing more nature produced increases in nature connectedness through the increased noticing. Complicated? Yes! As we've seen, the natural world is full of complex relationships, and we are part of them. But the message is more straightforward. When people take actions to create visible biodiversity, it boosts noticing nature which brings people closer to nature, which motivates actions for biodiversity!

There's a clear story emerging about the importance of nature connectedness for pro-nature behaviours – although we did see earlier that valuing and having concern for nature didn't explain those behaviours. This may seem odd, but the 'value-action gap' is a long-established topic when considering pro-environmental behaviours.[8] The role of nature connectedness or a person's emotional relationship with nature, the role of noticing nature, and wider work showing the role of emotions suggests a new focus for engaging people with pro-nature behaviours could be worthwhile.[9] That new focus could centre around increasing nature connectedness and closing the 'connectedness-action gap'.[10] It seems simple that people who are tuned into nature and have a close relationship with nature do more for it. Yet, one traditional aspect of nature engagement is improving environmental knowledge, and people still turn to knowledge, facts, figures and education to improve pro-environmental behaviours.[11] But this approach may be misguided – for example, millions of people may not be able to identify a sparrow,[12] but if they notice and enjoy it, they'll be more likely to care enough to learn its name. The inability to identify a sparrow is a symptom of the disconnect with nature. Treating the symptoms overlooks the root cause.

The meaning of *know*

Some of those seeking to create a more caring relationship with the natural world suggest something along the lines of 'We need to teach our children about the natural world so they learn to love it and therefore will fight to protect it'.[13] Some quote, cite or rejig Richard Louv: 'We cannot protect something we do not love, we cannot love what we do not know'.[14] Away from the further context that follows this sentence, the use of *know* could mean to 'be aware of through observation, inquiry, or information'.[15] However, the quote goes on, 'and we cannot know what we do not see. Or hear. Or sense' and the paragraph refers to 'attachment theory', which is all about deep and enduring emotional bonds. Louv continues to discuss sense of place, relationships, beauty and wonder. The second definition of *know* makes more sense: to 'have developed a relationship with'. The popular quote becomes 'we cannot love what we do not have a relationship with'. Thinking of person-to-person relationships, this makes complete sense. We could be taught the facts and figures of a stranger, but it's unlikely to lead to love and care. Yet to see, listen and sense that stranger, form attachment and emotional bonds with them, is more likely to lead to love and care. Love isn't taught.

The anthropologist Nurit Bird-David discusses how relational and modernist worldviews help explain the misplaced meaning of *know*.[16] The modern objectivist worldview sees learning as acquiring knowledge about things; the known and the 'knower' are separate, with the thing of interest being broken down into further separate parts in order to know it. Objectivist knowledge involves acquiring and improving representations of things. Whereas nowadays, a child at school or scientist may dissect a frog or flower, the animistic approach on the other hand involves a two-way interaction with the flower. Paying attention to how it responds to actions and the accompanying changes in oneself is a mutual responsibility – an understanding of relatedness from a related viewpoint. Animistic knowledge involves being in the world of other things, developing skills of awareness and relatedness – 'absorbing differences' rather than highlighting them. As Bird-David puts it, it's 'I relate, therefore I am' rather than 'I think, therefore I am'.

It is perhaps little surprise to find that recently nature connectedness has been linked with Indigenous knowledge and culture. Using a community-led approach, a strong positive relationship was found between nature connectedness and Anishinaabe culture of Central/Southern Canada. This relationship grew from communication through songs, stories and ceremonies, rather than being simply driven by repeated interactions with nature.[17] These findings suggest that the inclusion of Indigenous knowledge and culture more widely could help with the transformation of worldviews needed for a sustainable future.

It's pertinent that a book about knowing, *The Living Mountain*,[18] has become a classic of its genre. On the closing page, Nan Shepherd says 'Knowing another is endless … The thing to be known grows with the knowing.' The account in the book is full of observation, but observation that forms a deep relationship with the mountains of the Cairngorms. Here, 'Knowing another' suggests a relationship. So, for a love of and reconnection with nature, there is a need to be conscious of the meanings of *know*. To see, sense, hear, notice, experience, appreciate, feel, behold and be friends with the sparrow. As that relationship forms the bird's name, its ecology and cultural significance can be learnt.

Returning to our discussion of ways of thinking about nature, in *The Living Mountain*, Nan Shepherd demonstrated new thinking in line with philosophies being developed around that era by Merleau-Ponty, but without recognition at the time. The book is a study of consciousness and lived experience, a phenomenology that captures the very act of being – how our bodies are instruments of discovery that can be tuned to see the depth of being in nature. How we can find joy in the perception each sense can bring us as channels of existence in the natural world. At rest on the mountain, Shepherd found an uncompromised connection with nature, a place where the mind becomes uncoupled, and the boundaries of self dissolve to be at one with the natural world. Such classic accounts don't tell the story of facts and figures; they tell the story of emotions and relationships.

Unsurprisingly, despite many being keen on a teaching approach, research shows that focusing on education, information and knowledge is not a great route to pro-nature behaviours. A favourite study of mine for making the point is by Siegmar Otto and Pamela Pensini.[19] Together they looked into children's environmental education and the difference made to environmental behaviours. They also included nature connectedness, given the weak link between environmental knowledge and behaviour. A close relationship with nature provides the essential intrinsic motivation for adopting a more ecological lifestyle. Yet, although nature connectedness as a single construct has been found to outperform other variables as perhaps the strongest predictor of ecological behaviour, fostering nature connectedness is not a common feature of environmental education.

Otto and Pensini gathered data from 255 children aged 9 to 11, measuring ecological behaviour, environmental knowledge, nature connectedness and participation in environmental education. After careful checks, their analysis revealed environmental knowledge explained just 2% of the levels of ecological behaviour. In comparison, nature connectedness explained 69% – a stark difference. They also found that there's a benefit to teaching nature knowledge through a focus on nature connection. More recently, individual studies like that by Otto and Pensini were grouped and included in a meta-analysis of dozens of studies looking at the best routes to a more sustainable future.[20] Bringing together 147 studies, this research found that people with higher levels of nature connectedness demonstrated higher levels of pro-nature behaviours, but environmental education had a less positive influence.

When presenting such work and the need to develop an emotional bond with nature, the most common challenge that I receive is the argument that developing knowledge is the key. Our knowledge-based approach and current relationship with nature are deeply embedded. Perhaps thanks to Francis Bacon, we like to identify, name and classify nature to understand it. Scientific investigation and knowledge are important – I'm a scientist, and much of this book is based upon scientific investigation. But the evidence shows that an emotional relationship with nature matters, and we tend to favour teaching people to care rather than building a close emotional bond. Often the efforts to engage people with nature are based on identification and knowledge. We seem driven to know, understand, be smarter, walk further, run faster, climb, cross, conquer, consume and prosper.[21] Materialism and such pursuit and pace of life have been found to reduce people's level of pro-environmental behaviours.[22] Our attachment to things and the self is also problematic for pro-nature conservation behaviours. In our research, we've found that when people are free from the cycle of craving and attachment to things, they show higher levels of pro-nature actions.[23] The materialistic way of being contrasts to connecting with nature which, as we'll see in more detail later, involves less purposeful activities – simply noticing nature and the emotions evoked. These simple activities can develop into deeper explorations of the meaning of the natural world. Where science is about understanding nature, connection is nature better understood. This subtle distinction is important.

Towards one health

Rather than relationships and the unity of human and nature's wellbeing, attempts to increase pro-environmental behaviours tend to focus on the person or the context.[24] This contextual approach has in turn tended to focus on opportunities and barriers. At the personal and perhaps the simplest level, people can be provided with knowledge through information campaigns, but approaches also include egocentric motivations around self-enhancement, promoting social norms, group conformity, and motivations independent of human needs and based on the intrinsic value of nature. It's interesting to see how these approaches match people's reasons for protecting nature and how they vary across countries and cultures. Is the aim to ensure our health and survival through protecting resources? Or are more altruistic and ecocentric motives at play, such as preventing declines in wildlife and maintaining the interconnected ecosystem we are part of?

As you'd expect, these reasons differ across countries and cultures. Using responses from 12,000 interviews across 11 countries, Abel Gustafson and colleagues looked for answers to these questions.[25] By looking for themes, they found human health and survival were the most frequent reasons people gave for protecting nature, present in 22% of all responses. This human-centred and utilitarian response was most common in China, Brazil, South Korea and South Africa at around 30%. However, there were some quite significant differences in motives. In India, only

7% emphasised human survival, with 42% focusing on environmental health. Indeed, India had by far the most nature-centred responses, with nature being mentioned in 72% of reasons. This compares to around 38% in the USA and China. The UK was a little more nature-centred, with 44% of responses including altruistic motivations. Flip the stats around, and the USA had 44% and China 54% human-centred responses – how philosophies have changed since the Chinese Song dynasty. Intriguingly the UK wasn't far above India's 20% human-centred reasoning, at 25%. The UK gave more responses around protecting nature for future generations of people or moral imperatives; however, the researchers didn't classify these as either human-centred or nature-centred.

These perspectives are interesting; however, although India has achieved notable successes, it also has nature conservation failings.[26] Still, the potential source of these ecocentric perspectives is interesting given that India perhaps had the earliest wildlife protection laws, in the third century during the time of Emperor Ashoka who devoted himself to the Buddhist concept of Dharma.[27] In South Asian and Indian religions, Dharma is an ethos of central importance and has been related to ecology in Hindu communities in India[28] – for instance, mass tree planting of the Svadhyaya movement, and the tragic incident of 363 people of the Bishnoi community losing their lives hugging and protecting a stand of trees in 1730.[29] The Bishnois were early proponents of wildlife protection and married eco-conservation with their faith.[30] Dharma has a deep interface with ecology, which includes humans in a global family and unites human and nature's wellbeing.[31]

As we saw earlier in the book, popular models of health and wellbeing overlook nature, but recently more Dharma-like 'one health' models have been proposed, discussed, endorsed and outlined in medical journals.[32] These models recognise that humans are embedded within the rest of the natural world and that a fundamental factor in wellbeing is a healthy planet. They stress the connections between environmental, animal, wildlife and human health, from a molecular to planetary scale. These models reflect a broader and dynamic interconnectedness, a 'biopsychophysis', where health depends on biological, psychological and natural environment factors.[33] There are a number of holistic approaches such as 'ecohealth' and 'planetary health'.[34] These concepts have much in common, yet some note the differences and feel the distinctions are important. However, to my mind, a new relationship with nature requires one model that shows how biodiversity and human health are connected.

In 2010, Paul Stevens presented a new paradigm for wellbeing based on an ecopsychological view of health.[35] This fascinating account describes the many ways that we as humans are interconnected and embedded within the environment via two-way electromagnetic, chemical and mechanical interactions. Everything we know of the world comes through these interactions, and they impact our wellbeing, yet our concept of a separate 'self' and 'environment' remain rigidly fixed. This is despite the presence of our bodies extending beyond the skin-defined boundary, be it through breath or ions moving through cell walls.[36] When our environment changes, these interactions change and our wellbeing changes.

Models of health that reflect the reality that people are embedded within the natural environment provide a shift of emphasis away from focusing on the person and their health issues to considering the relationship between people and nature. If these models were widespread, they'd provide a better understanding of the importance of nature for wellbeing, which would lead to greater respect for the natural world. Recognising these relationships and a one-health perspective can inform and unite activity from health services to environmental policy and beyond, to unify human and nature's wellbeing. If promoted as the core approach to wellbeing, it would emphasise the importance of the human–nature relationship to people during their everyday lives – an essential step to a more sustainable future. Thankfully, global institutions such as the UN are recognising that the health of people, wildlife and the wider environment are linked, and are looking to advance one health and restore the 'broken relationship with nature'.[37]

So far, we've considered the broken relationship with nature and the benefits to human and nature's wellbeing that restoring that relationship can bring. I hope a strong case has been made for nature connectedness providing a useful focus and a route to a sustainable future. So, we'll move on to looking at the ways in which nature connectedness has been improved to help bring about a reconnection with nature.

PART III

CREATING
A NEW RELATIONSHIP
WITH NATURE

CHAPTER 10

The Good Things in Nature

First steps

Moving from a broken relationship with nature to the benefits of reconnection is a good thing. We need ways to improve the human–nature relationship, which can be done by targeting nature connectedness. In some ways, what follows in this chapter may seem rather straightforward given the entrenched disconnection of most of us humans from the rest of nature. However, there has to be a starting point, and the final part of this book will move from simple interventions and scale up to ideas around the societal change needed for a sustainable future.

As we saw in the preface, the starting point is provided by the initial steps I took, quite literally wandering into a closer relationship with nature. In 2012, I stopped writing on foot and began researching. Rather than starting by seeking to understand the psychology of nature connectedness, convinced by my experience and the benefits of reconnecting with nature, I set out to find ways to improve matters that didn't require 500 walks and 90,000 words. I started planning a task, distilling the approach I'd stumbled on into its simplest form. Simplicity was important, as previous work on improving the human–nature relationship had involved wilderness experiences.[1] These are time-consuming, impractical for many and difficult to embed into daily routines. In modern times, a new relationship with nature must be everyday and, for many, urban.

It was known that brief exposure to nature could lead to short-term increases in nature connectedness, and the challenge of delivering sustained improvements had been identified.[2] My inspiration came from brief positive psychology interventions, or PPIs. Regularly counting blessings had been found to improve happiness and wellbeing, and this included simply writing down three good things each day.[3] However, as we've seen, most people don't notice nature. So the likelihood was that people would not choose to write about the good things in nature. Indeed, we found that when asked to write three good things each day, people didn't write about nature. People wrote about work and relationships, and their stress levels actually increased – perhaps because they were revisiting issues they'd encountered during the day while finding a good thing, which often referred to their job – for example, standing up for themselves at work. We couldn't find a psychology journal that wanted to publish this negative result, so after six years, we published it independently.[4]

There was a need to prompt people to notice the good things in nature. A simple twist was to add the words 'in nature', to create a task where people wrote about the good things in nature. Prompting simple moments to notice the good in everyday nature, from the clifftop to the bus stop. The guidance people received before taking part was informed by an analysis of my writing in 2010.[5] This revealed a move from an identification and knowledge-based relationship to a sensory and emotional relationship where the beauty of nature was important. The new nature connection intervention asked people to 'write three good things in nature that you noticed today', with further guidance that read, 'Things you can list can be the beauty of small things at any one moment or wider aspects that arise from attending to the diversity and wonder of the natural world around you. For instance, it could be as seemingly trivial as noticing the song of a robin or movement of a tree in the breeze.' In 2013 we finally ran the study, asking 50 people to write three sentences about the good things in nature each day for five days. A control group of 42 people noted three factual things, such as what they had to eat. As mentioned above, there was also an original 'three good things' group. As ever, we took measures before, straight after and then again two months later. We found that the nature group benefited from sustained and significant increases in nature connectedness compared to the control group. The increase in nature connectedness also helped explain improvements in psychological health.

Research completed, later in 2014 we submitted the paper to a journal for publication. That year saw eleven sets of reviewers on three revised versions of the paper, with several positive comments. However, the editor decided not to publish the paper – an insight into the worst aspects of research. Two other journals didn't see value in an intervention to increase nature connectedness; another sign of the general disconnect. It wasn't until November 2016 that the study was finally published, another nine months after submission, but still perhaps the first intervention to show sustained improvements in nature connectedness.[6] The research has gone on to be cited by dozens of other researchers, and, as we'll see, further research has shown how effective the intervention is. More and more journals are now publishing research in this area, with the number of mentions of nature connectedness increasing nearly eight-fold from 2014 to 2022.

It's worth thinking about how simply writing down three good things in nature for five days could have an impact two months later. When people know that three good things are required each day, they are more likely to be noticing nature during the day; the task goes beyond the time taken to write a few sentences. The short amount of time spent writing does not reflect the time spent thinking and reflecting on moments in nature. As we've seen, conscious self-directed reflective thought explains nature connectedness more than mindful attention. The act of writing, even if it is only for a few minutes each day, is not simply a substitute for speech or output of thought; it can shape thinking.[7] Sentences carry thought and help create new thinking. In this case, the writing develops a person's thinking about their relationship to nature, and this process involves the natural world around us.

We've seen how everything is related, and this includes our thinking, which works with the wider environment as a 'coupled system'.[8] From this perspective, what we notice alters our thinking, as does what we write. So, the impact of noticing nature is reinforced by the writing; the act is weaving us back into the natural world. This view of thinking being integrated with the wider world has its roots in phenomenology and the work of Merleau-Ponty which, as covered earlier, emphasises humans' shared place in the natural world.[9] It's a perspective supported by observations that Indigenous peoples often perceive nature to be integrated within their own self.[10] This perspective highlights how our thinking is embodied and extends beyond the head, embedded as we are in the wider environment as biological beings that evolved to fit and understand the natural world.[11] A simple task can trigger powerful mechanisms.

The process of being woven back into nature such that it merges with our sense of self can be seen in my reflections from 2011:

> The evening calmed to an oblique blue and time
> had stolen July. I listened to the blackbird's final
> boast of summer, its song a twine for thoughts
> to climb and board a vessel of reflection, riding
> an ocean of greens, canopy hues that splash
> the air of sky and deny it'll ever be autumn.
> Aboard and skyward, my sail was weaved by
> a weft swift, through warp jute threads on the
> sky loom. Each note of the song a stitch, every
> flourish a tie to capture the breath of summer
> and journey to a place where trees puddled the
> fine dusk horizon, pools of blackness against a
> trace of setting sun. I felt alive in the stillness,
> as a star is bright in the night sky. Where the
> river was audibly unstill I looked out over the
> flatlands, from foreground to far, and felt that
> the landscape and my mind merged, my sense
> of self dissolved.

More good things in nature

We have seen how nature-based solutions are often overlooked in wellbeing guidance. Part of the reason is perhaps the lack of access to straightforward options. Hence, the next step with the 'three good things in nature' approach was to make it accessible.

Smartphones are an accessible item for many. Technology may be associated with a more distant relationship with nature, but it is here to stay, and it's how it is used that matters; even the simple technology of a pen can be used in positive and negative ways. As part of the 'Improving Wellbeing through Urban Nature' project, which ran from 2016 to 2019, we developed a smartphone app that included the 'noticing the good things in nature' task. We know that people don't tend to notice nature, so this approach allowed people to be prompted to do so each day for seven days.

The app tracked the user's location and use of green spaces in Sheffield, using about 1,000 'geofences'. As most people spend the vast majority of their day indoors, they were reminded to notice the good things in nature whenever they approached a green space. The developers we worked with were terrific, but the memories I have of testing the app and wandering around Sheffield waiting for an alert are vivid.[12] To help engagement further still, people had a 'conversation' with the app via a 'chatbot' that kept emotions positive and made engagement with nature easy.

This study was much larger than the first, with 582 adults noticing the good things in nature or in built spaces. This included 148 adults with a common mental health problem. Again, we found significant and sustained increases in nature connectedness. The 25% increase in nature connectedness explained significant and sustained improvements in mental health. The mental health improvements were clinically significant for those living with a mental health difficulty.[13] As well as providing evidence that noticing the good things in nature has clinical potential as a wellbeing intervention or social prescription, the research also showed the causal link between improved nature connectedness and improved mental health.

As you may imagine, we found the benefits were greater for those who had spent less time outdoors in the last year and started the study with lower levels of nature connectedness. A smartphone app is likely a good way to engage those who spend little time outside with nature, especially the everyday urban nature close to where they live. We also found that the people who reported that they'd spent more time outdoors as a child improved more too. We've seen that childhood nature connection drops sharply in adolescence, and this perhaps suggests a 'latent' relationship with nature that was reignited by noticing the good things in nature.

Earlier in the book, we considered how nature can help manage our moods, how the dimensions of the three-circle model help explain the ways in which engaging with nature can bring the joy and calm that can help balance our emotions to keep us well. A further part of the Sheffield research included a multidimensional measure of positive emotions.[14] This allowed us to distinguish between relaxed positive emotions and high-arousal positive emotions. We found that noticing the good things in nature increased relaxed positive affect, and this also helped explain the benefits to mental wellbeing. These low-arousal, calm, positive emotions are linked with benefits to life satisfaction, mindfulness, and reduced depression, anxiety and stress – to a greater extent than high-arousal positive emotions.[15] This also supports accounts about how improving nature connectedness brings mental wellbeing. Noticing and noting the good things in nature helps people manage their moods in the same way as being in woodland, viewing flowers or touching wood.

Three good things in nature for mental health

We had perhaps the first empirical evidence that targeting nature connectedness to create a closer relationship with nature could bring sustained and clinically relevant improvements in mental health. Simply, prompting engagement with urban nature each day can improve nature connection and wellbeing. This can be done in a low cost and widely accessible way, providing solutions to mental health issues in increasingly urbanised populations. Indeed, the successful RSPB nature prescription pilot in Edinburgh was focused on nature connection and included noticing three good things in nature.[16] In the pilot, five GP practices prescribed nature to 350 patients, and the results revealed that 91% of prescribers would continue to prescribe it, and 87% of patients would continue to use nature for wellbeing. Feedback showed that patients felt they had 'permission' to engage with nature, and it was the sensory and emotional connection that most patients enjoyed and benefited from.

Combining prompts to engage with nature with walking appears to be effective. In related work, people who were asked to 'tap into their sense of wonder' and take selfies on eight weekly walks showed a number of benefits compared to a control group.[17] Participants reported more positive emotions, more joy and felt part of something larger. The photographs revealed greater smile intensity and a 'smaller self' through focusing more on nature and less on themselves. Our research team has also combined the 'three good things in nature' approach with walking. This time, noticing three good things in nature was combined with a nature-based walk in small groups.[18] This is an approach that could work well with social prescribing for people living with the common mental health conditions of depression and anxiety. Fifty people took part, and most had a formal diagnosis of depression or anxiety, with the rest having symptoms. Most were either taking medication, accessing primary care support, accessing community mental health services under the care of a psychiatrist, or had been discharged but wanted wellbeing support. They were randomly allocated to an urban control or nature walking group that used the 'three good things in nature' approach. The guided walks lasted for 30 minutes on five consecutive days in groups of up to ten. The nature walks varied each day and took place in various natural landscapes. As you'll be expecting, measures of nature connectedness and mental wellbeing were taken at baseline, after a week and a six-week follow-up.

Once again, the analysis showed a significant and sustained increase in nature connectedness in the nature versus the urban walk, with the increase at follow-up reaching 30%. So, what did focusing on nature connectedness mean for mental health? Well, the people in the nature group also reported significantly higher levels of wellbeing at the six-week follow-up compared to the urban group. The 'three good things in nature' intervention worked again, with this study showing the benefits in a clinically relevant population – meaning this cost-effective and straight-forward approach could work well for those living with low-level depression or anxiety.

We saw earlier how large-scale survey work showed that a closer relationship with nature and simply noticing nature explained wellbeing. This repeated empirical work provides evidence of a causal link between nature connectedness and mental wellbeing. Moreover, for context, the 10% and over improvements in wellbeing at follow-up can be compared to other, often more time consuming and accepted, approaches. For instance, an evaluation of three digital cognitive behavioural therapy (CBT) approaches found increases of between 9 and 14%,[19] with other CBT-type approaches, such as guided imagery, finding benefits of 5%, 7% and 13%.[20] Away from CBT and towards mindfulness, an intervention involving nine 75-minute sessions, plus additional home practice, found a benefit of 15% at the end of the programme.[21] Turning to other nature-based approaches, a six-week programme of nature walks and activities such as birdwatching found an improvement of 11%.[22] Combining three good things in nature with walking worked particularly well, and of course, the nature connectedness focus would likely be improving pro-nature behaviours at the same time.

Indeed, we included pro-nature conservation behaviours in another study involving around 500 people. Here, the main aim was to explore another approach to nature engagement, citizen science, alongside the three good things in nature. You'll be familiar with the approach now: people were allocated to a short task and completed it each day for five days with measures before and after. The ten-minute tasks included three good things in nature, as well as citizen science activities: counting insects visiting a flower, and a butterfly survey. Another group did nothing. As we've found before, noting three good things in nature increased nature connectedness and wellbeing measures. So did the citizen science tasks, although a little less markedly for nature connectedness and for wellbeing. However, it was only noting three good things in nature that had a significant effect on pro-nature conservation behaviours – a little surprising when the citizen science tasks were related to nature conservation, and participants felt they were helping care for nature. Targeting the human–nature relationship is clearly vital for pro-nature actions.[23]

What are the good things in nature?

You can imagine that during this research we amassed thousands of sentences about the good things in nature! So it's interesting to look at what people wrote about. Knowing what people view as the positive elements of everyday nature can help develop new ways to engage people with the natural world. So, we have analysed many of the sentences, looking at the common themes that emerge. These themes help us understand the routes to a closer relationship with nature. One key finding is that people know what nature is. Sometimes researchers get bogged down in the need to define nature, asking for example if it includes animals or pets, or arguing that human constructions and buildings are nature, just like a spider's web. I don't get involved in these contortions and cul-de-sacs – and what we found is that people overwhelmingly don't write about pets and buildings, with farm animals not featuring a great deal either.

Our analysis of the first 1,000 good things in nature revealed ten themes, five of which accounted for over 70% of the good things in everyday, often urban, landscapes.[24] Let's look at those key five themes in turn. The sensory experience of nature is an essential first step of access to the world, and people wrote about the sights, sounds and smells: the singing birds, crashing waves and the scent of flowers. The way nature changes over time was also a popular topic – the buds of spring, new leaves, emerging blooms and the turn of the seasons. People also enjoyed active wildlife, such as the simple movements of birds or squirrels chasing about the park. The effect of the weather often featured too: the breeze in the trees or raindrops glistening on fallen leaves. Finally, nature's beauty and wonder were also seen as good things, from distant views to the wonder of a spider's web.

We performed a similar analysis of the good things in nature from the Sheffield study with a different team of researchers.[25] Similarly, one dominant theme that emerged was people's wonder at encountering wildlife, including the sensory aspect of hearing birdsong. The second largest theme identified was general gratitude for street trees, a salient feature and controversial topic at the time given the felling of many.[26] Sheffield is also a hilly city, and the third most common theme was the awe participants expressed at views from higher ground and looking down over the city, often under dramatic skies. Flowers, water and nature's beauty came through once more too. Overall, biotic themes – that's living organisms – had greater representation than abiotic themes such as skies and views.

Growth, change, active wildlife and trees are beautiful things to engage with. Simple things, yet meaningful. Noticing them and writing them down can bring us closer to nature, and that's good for our mental wellbeing. When we find our place in nature, life makes sense. No wonder two years and 90,000 words had a profound impact on my relationship with nature. It's interesting that the 'three good things in nature' approach had a greater impact when combined with walking. Walking moves us through the landscape, bringing new perspectives. When writing on foot, there is a need to break the rhythm of walking, to walk until the landscape suggests a place to stand. To pause, listen, reflect.

I found that my notes were often brief, simple cues so as not to break the present moment. But they were enough to prompt revisiting the moment later in the day to turn it into prose. Yet, as time progressed, short phrases would come to me and became the catalyst and theme for particular walks. Starting a flow of words that would continue and then stop as quickly as it had emerged. Leaving accounts often full of growth, change, activity and nature's beauty. A library of good things I can now access and use to travel in time, to a moment a decade or more before, to find:

Change:
From their distant perch, hidden rooks bled their calls into the air of changing season.
The landscape was at its most bare and bleak, where I could become the flesh to its bones.

Calm:
I followed the river to the shallows where it spoke of its bed, in whispering tones as gentle as the sun's growing warmth.
Patient in a contentment of place, I watched the heron's silence.

Wonder:
Swifts scrape the sky, their blades slicing its limits, fragments of sound falling like sparks from punctures that will become stars.
At dusk, a blackbird perched high, its song the essence of the day being imbibed into its blackness for release at first light tomorrow.

Connection:
I walked the wet sand that resists like bone, then gives like flesh.

And as connection deepens, loss:
The cuckoo sounds over the flatlands by the river where its call can travel the void left by those lost – its spring repeat now a reminder of decline.
The landscape rose to a place where birch cried autumn and ash trees stood like ghosts of their future selves.

Nature always has a story to tell. And the good things in nature help keep you well.

CHAPTER 11

Pathways to Reconnection

Noticing nature

Noting three good things in nature each day is just one intervention to improve the human–nature relationship. Given the scale of our broken relationship with nature, many more approaches are needed. As we've seen, people tend not to notice nature and so a prompt to do so through noting the good things in nature has encouraging benefits. In another study, we explored how important noticing nature is for nature connectedness. We also wanted to explore how the active sensory engagement of noticing compares to more general nature contact.

Using a survey of over 2,000 adults in the UK, we asked questions about noticing nature activities such as watching wildlife, smelling flowers and listening to birdsong, and frequency of contact with nature (e.g. walks and park visits). We then looked at how these factors explained levels of nature connectedness. We found that noticing nature explained nature connectedness over and above merely spending time in it, with simple 'noticing nature' activities accounting for around 50% more of the variation in nature connectedness than nature visits. Clearly, to notice nature requires time in nature, and we found that noticing and contact with nature worked together to explain over 60% of the differences in levels of nature connectedness.[1] There is a great deal of work ongoing that aims to increase access to nature and contact with nature. Encouraging visits and time *in* nature is a good thing. However, a trip to the park is just a first step. For a new relationship and the benefits that brings, there's a need for engagement. There's a need to spend time *with* nature. As when stepping into a room of strangers, engagement is needed to form new relationships. Focusing on access alone isn't going to fix the broken relationship with nature. There's a need to want to engage. And to know how to engage.

The survey above was cross-sectional, informing us of what factors explain levels of nature connectedness. The empirical work around noticing nature helps to show the causal link. In addition to our 'three good things in nature' approach, others have focused on noticing nature and the importance of prompting people to engage with nature in simple ways. In an Israeli study, people receiving prompts to smell and touch nature interacted with nature more and experienced positive wellbeing and connectedness benefits. People who were simply asked to walk slowly or turn off their phones didn't benefit.[2] However, follow-up measures weren't taken to look for sustained benefits.

We can also return to Canada, where, coincidentally around the same time as we were noticing the good things in nature, Holli-Anne Passmore asked people to pay attention to how nature made them feel and take photographs of those things that evoked emotions. They were also asked to write a description of those emotions. This was each day for two weeks. Compared to the control groups, those in the nature group had significant improvements in nature connectedness, wellbeing and, interestingly, pro-social behaviour. The research also found that the increases in connectedness were over and above those from contact with nature, even though people spent equal amounts of time in nature.[3] In an extended replication of that approach, Chinese university students who took part also had less ill-being and greater wellbeing. Similarly, the benefits were unrelated to time spent in nature, and instead the benefits were from noticing everyday nature in their daily routines.[4]

Noticing nature approaches have also been taken by Alexia Barrable – through asking students on a single ten-minute urban walk to firstly 'notice the beauty in nature', and then either simply 'mentally note' three beautiful things or use photo, video or sound recording on their phone. Both approaches had a significant effect on nature connection, with the second bringing slightly larger benefits of around 10%.[5] In a second study, children were asked to listen to nature sounds or notice sights of nature such as flowers. The emphasis was on mindful or active attention in the here and now.[6] Time spent engaging in these activities led to short-term increases in nature connectedness of around 33%.[7]

The benefits of noticing nature are becoming clear. However, we've already seen that people spend little time outside in green spaces, and those with a more distant relationship with nature tend to look at buildings rather than the natural world.[8] In contrast, people who have a closer relationship with nature are better at noticing it, having enhanced attention capacity[9] and tending to explore natural scenes more,[10] reinforcing that relationship. This preference seems to be driven by seeing a relationship with nature as an essential part of one's self, rather than straightforward enjoyment of nature experiences. There's a need to break out of a cycle of not noticing nature, but there's not a great deal of research in this area, and as we've seen, there's also an ongoing battle for attention and less nature around to notice.

As well as ways to break the cycle and prompt people to notice nature, another area for further work is finding ways to bring the sustained improvements in nature connectedness needed for an enduring close relationship. There are dozens of studies that have brought short-term boosts to nature connectedness. In a review of nature connectedness interventions we found 12 that included the follow-up measures needed to show sustained improvements.[11] The good news is that together those studies confirmed that sustained improvements in nature connectedness are possible. The sustained benefits involve a variety of forms of engagement, but typically involve being prompted to engage with nature regularly.

Challenging assumptions

Getting people outside into nature and the use of knowledge-based activities are common engagement approaches used to reconnect people with the natural world. Shock and national headlines often abound at the population's lack of nature knowledge and inability to identify wildlife and plants.[12] This is likely another symptom of our disconnect with nature, but that does not mean treating symptoms is a route to a closer relationship. What types of lasting relationships are based on facts and figures?

Another assumption is that the solution lies in simply being outside in nature. This often gives a short-term boost to nature connectedness, but there are mixed results, with contact with green spaces such as a vegetated courtyard or parkland not always leading to these short-term increases.[13] In a wider survey no relationship was found between nature connectedness and nature exposure, visit frequency and nearby green spaces.[14] Taking nature visits up a notch, Ian Williams and colleagues explored the impact of intensive seven-day outdoor adventure experiences on adolescents in Australia.[15] They put a great deal of effort into addressing many of the limitations apparent in previous evaluations of outdoor programmes through adopting more robust methods, designs and measurement tools. However, they didn't find any impact over the short term or medium term on 14 of the 16 measures, which included nature connectedness. It seems likely that this is because of the focus on challenge and adventure. Elsewhere, less prescriptive outdoor adventure courses based on individuals exploring the natural environment have positively impacted people's relationship with nature. However, these results were based on interviews rather than measurement.[16] More recently, environmental education that included a focus on emotions and sensory contact has been effective in increasing nature connectedness in the short term.[17]

We've seen that increasing knowledge is not a route to increasing pro-nature behaviours. There's also research that shows that the knowledge-based approach is not an effective way to bring a sustained increase in nature connectedness. When comprehensive environmental education programmes that focus on knowledge are evaluated, they struggle to show these longer-term effects.[18] For example, working with a few hundred university students in the USA, Greg Lankenau taught an introduction to ecology course in two different ways while keeping the course content the same. For some, an environmental science perspective emphasising knowledge acquisition was used. For others, the emphasis was on celebrating connectedness and relationships. A traditional geography course was also included. Only those students taught with a focus on connections and relationships with nature had enhanced nature connectedness.[19]

Similarly, Elisabeth Nisbet found that Canadian university students studying courses such as ecology, environmental sciences, geography or natural history showed no increase in nature connectedness over the term.[20] Other knowledge-based methods have been compared to approaches such as art projects. For instance, Coral Bruni compared a knowledge-based quiz trail to art-based activities where people used meaningful nature-based resources. It was the art approach that increased nature connectedness.[21]

However, these were only short-term effects, and when long-term follow-ups are included, the short-term effects can soon disappear.[22] There are a couple of studies that, like 'three good things in nature', have managed to bring about a sustained increase in nature connectedness.[23] Several hundred young and teenage children in Singapore took part in a range of outdoor activities over one or five days. This included a 'hands-on' exploration of the local natural environment and learning about natural heritage and ecosystem ecology. The five-day programme worked best and this went beyond contact and learning to emphasising 'memorable' and 'positive experiences' with 'emotional excitation' and improved 'affective response'. Similarly, in Germany, a hands-on approach, quite literally with activities such as using touch to recognise trees, produced sustained increases in nature connectedness.[24]

However, overall, when combining the results of several environmental education studies, a meta-analysis found no significant impact on nature connectedness.[25] The explanation for this was that the traditional approach of transmitting scientific knowledge in an anthropocentric and rational way suppresses emotional content. When children are taught to understand through maths and formulae, the measurable becomes primary. Emotions and relationships, what it is to be a human as part of nature, becomes secondary. That which can be ranked, indexed and measured becomes the reality.[26] Green becomes #00FF00 in hex code. Yet transferring information and knowledge is the focus of many attempts to connect children with nature.[27] In sum, our focus needs to shift from delivering knowledge to building relationships.

Arjen Buijs and Anna Lawrence from the Netherlands highlight how the tendency to 'rationalise' nature has 'pushed emotion out of sight'.[28] They note how emotions are labelled 'irrational', a history we explored earlier when considering the roots of our broken relationship with nature. They also highlight that those emotions influence our thinking and motivate change, so it's no surprise that they call for research to incorporate emotions and state that emotions need to be visible.

Together, these studies suggest that getting outside and being educated are not effective routes to fostering a closer relationship with nature. Furthermore, although it's a great start, focusing on noticing nature alone won't spark enough reconnection with nature across society; with our current tendency to destroy nature, there's a need for transformational change. People need various pathways to nature, and there's a need for evidence-based frameworks that go beyond physical access, to emotional access and the creation of cultural pathways to normalise a close relationship with nature – pathways to a nature-connected society.

The five pathways to nature connectedness

The good news is that such a framework exists. Let's consider the work that went into developing it. At the start of the book, when considering the broken relationship with nature, we saw that Stephen Kellert identified nine types of human–nature relationships. Four of these were unrelated to a close relationship with nature.[29] These were fear of nature, dominion over nature, utilitarian use of nature and a purely scientific relationship.[30] We can now turn to the other five types, those related to a closer relationship with nature. These include aspects touched on above in the successful approaches, such as creating emotional and meaningful experiences. We identified these specific routes to nature connectedness in a three-year research project that produced a widely adopted framework for those seeking to reconnect people with nature.

Ryan Lumber led this research as part of a PhD I supervised alongside David Sheffield. The research started with two surveys of how people engaged with nature activities, defined by each of Kellert's nine types of human–nature relationships. As we've seen, more knowledge-based activities, such as understanding and observing nature, did not explain levels of nature connectedness. Similarly, activities based on the use of nature, simple things like growing vegetables and hunting, did not explain them either. Furthermore, representing the control of nature, activities such as using natural spaces for sport were unrelated to nature connectedness. These are results that some find challenging, so they were repeated to check. However, in reality, people do not pursue activities in nature with laser-like focus. Tending the vegetable patch or identifying birds will involve other types of relationships with nature. They will involve the senses, emotions, creating something meaningful, perhaps bringing about a sense of care and compassion for the beauty of the natural world. This is what we found. The two surveys showed that sensory contact, emotion, beauty, meaning and compassion explained levels of nature connectedness.[31] A closer, healthier and more sustainable relationship with nature comes via these aspects – just like human-to-human relationships.

In a third study, the five pathways were brought to life within a 'walking intervention'. During a walk, people took part in activities such as having a conversation about their feelings about the nature around them, and writing down the meaning of the nature they saw. Taking part led to a significant increase in nature connectedness when compared to simply walking in nature – showing again that simple exposure isn't enough. The pathways to nature connectedness framework was thus established and the guidance has been refined to:

> Senses: Noticing and actively engaging with nature through the senses. Simply listening to birdsong, smelling wildflowers, or watching the breeze in the trees.

> Emotion: Engaging emotionally with nature. Simply noticing the good things in nature, experiencing the joy and calm they can bring, and sharing feelings about nature with others.

Beauty: Finding beauty in the natural world.
Simply taking time to appreciate beauty in
nature and engaging with it through art, music,
or words.

Meaning: Exploring and expressing how nature
brings meaning to life. Simply exploring how
nature appears in songs and stories, poems and
art, or by celebrating the signs and cycles of
nature.

Compassion: Caring for nature. Simply thinking
about what we can do for nature and taking
actions that are good for nature, such as creating
homes for nature, supporting conservation
charities and rethinking our shopping habits.[32]

Thinking back to the 'noticing the good things in nature' approach, it is easy to see how it activates these pathways to nature connectedness. Clearly, sensory engagement is an essential first step – turning hearing into listening, for instance. Such noticing leads us towards activating the other pathways: appreciating beauty, feeling emotions, making meaning and displaying compassion.

Sensory contact with nature seems so easy – yet as we have seen, nature is a loser in the battle for attention. People tend not to notice nature; our attention is often attracted elsewhere, or nature merely provides a background for activity rather than being the focus. So there's a need to prompt and engage the senses, be it through smartphone alerts, art installations or, as we'll consider later, using the power of affordances through clever design. This issue of attention matters, as the rest of the pathways are built upon tuning in through the senses; after all, they evolved to make sense of the natural world.

The importance of emotion will come as no surprise. We've seen how sensing nature helps manage our moods, how biodiversity brings greater positive emotions and that the essence of a connection with nature is emotional. Yet we have also seen our divorce from emotions, how they are pushed out of sight even when trying to engage people with nature. This is a situation founded on the divide between emotion and reason, where truth comes from repressing emotions. Yet all along, those pursuing wealth have understood the power of emotions and how they drive attention. The beauty of nature, the simple form of flowers, becomes emotion through the senses. Indeed, perhaps it's more meaningful to say that flowers are what emotions look like, or birdsong is what emotions sound like – such that nature's beauty promotes connection with nature across cultures.[33] That's a story reflected in many of the greatest works of poetry and art, the art that both reflects and brings meaning to our lives. From Keats' 'Ode to the Nightingale' to Wordsworth's 'Daffodils', nature means a great deal to humans.

Gregory Bateson noted that art is a search for meaning and that finding meaning involves a form of thinking that is most in tune with nature.[34] Meaning is a broad and powerful pathway that can include the properties of a particular place in nature, the symbolism of nature, and it is meaning that can unite the other pathways. It is a deeply meaningful and emotional relationship with nature that appreciates its beauty, that builds a moral duty to treat nature with compassion, undermining self-focused beliefs and fostering other-centredness.[35]

We have already seen the evidence that there is a need to move beyond focusing on knowledge and study of nature, and a need to move beyond superficial contact with the natural world. Having said that, of course, knowledge-based activities should not be abandoned – an understanding of nature is essential to protect and inform nature's recovery. However, when attempting to develop a meaningful relationship with nature, the delivery of knowledge must activate the pathways to nature connectedness.

The pathways provide a broad framework for the design of creative engagement and alternative values and frames to the traditional routes of knowledge and identification often used when engaging people with nature. Providing facts, figures and labels is an easy option, but we've seen how the journey from knowing to caring is not about these things. To prompt the noticing of nature, its beauty and the emotions and meaning it brings is more challenging. Still, that challenge is at the heart of creating a new relationship with nature, one that displays compassion and care.

Applying the pathways

Thankfully, many creative individuals want to reconnect people with nature, and some have adopted the pathways – using the framework to prompt new ways to engage people with the natural world. Their openness to new ideas and creative thinking has led to this research into nature connectedness receiving national recognition from the Green Gown Awards, which recognise exceptional sustainability initiatives being undertaken by universities. Similarly, the work was named by Universities UK as one of the UK's 100 best breakthroughs for its impact. Many organisations and projects are using the pathways, from local to international; let's just consider two.

The first use of the pathways came in 2015, while they were still being developed. We were talking to the Wildlife Trusts about their new campaign, 30 Days Wild. This centred on a menu of 'random acts of wildness', and the pathways were used to review the long list of activities. The knowledge-based activities were highlighted for revision, and suggestions were made where an activity could be tweaked to activate a pathway, by noticing the beauty of nature, for example. 30 Days Wild has been a great success, with over a million people participating. Three evaluations and a five-year review show how taking part in the activities over 30 days led to sustained increases in nature connectedness, wellbeing and nature conservation behaviours.[36]

The National Trust has also used the pathways to design visitor experience. It was fascinating to work closely with them to help roll the pathways out to various teams within the organisation. Their excellent staff already had great expertise in engaging people with nature, but were still enthusiastic and creative in developing new approaches to creating a closer relationship with nature for their members and visitors. It was wonderful to see how working together, the pathways can help tweak existing activities, inspire new ideas or give 'licence to talk about emotions'.

The use of the pathways with the National Trust included a refresh of their '50 Things to Do Before You're 11¾' campaign. This work provides simple examples of how applying the pathways can suggest simple changes. For example, the slightly dominionistic 'Climb a tree' became 'Get to know a tree' – here children can still climb a tree, but a broad range of other activities can be included to help develop a more meaningful relationship with nature. Similarly, the purposeful 'Canoe down a river' became 'Float in a boat'; this could involve a journey or simply taking a moment to drift, notice and find joy or calm. Finally, the world of binoculars and identification suggested by 'Go birdwatching' was reframed by simply suggesting children 'Watch a bird'.

Although simple, these changes can be powerful, as we saw earlier in research that prompted people to 'tap into their sense of wonder' when walking. We've seen the links between a closer relationship with nature and wellbeing, so the pathways can be used to inform mental health interventions and programmes too. To explore this, we ran a research project that tweaked 'Watch a bird' further, simply activating the emotions pathway by prompting people to watch birds for joy.[37] Over 150 people took part in this 'joy-watching' study. Some counted bird species in their gardens for 30 minutes, while others, the joy group, were asked to rate their feelings of joy on seeing the different species. Noticing nature is good for us, so it's no surprise that people in both groups had improved wellbeing, decreased anxiety and increased connection to nature after taking part. However, those in the 'joy' group had a larger decrease in anxiety, with levels dropping by over 20%. Focusing on the emotion pathway can enhance the benefits gained from watching birds. This is encouraging as it's a simple activity that many people can do from home or locally.

Of course, we've seen how people don't tend to engage with nature and that nature is missing from our models of wellbeing. So to attract the disconnected people who can benefit most, joy-watching birds could be a structured activity used in green prescription schemes or adopted by schools and workplaces using 'bird therapy' stations. However, we've seen the complex relationships in nature and ecology; before doing such things at scale, there would be a need to consider unintended consequences and the imbalance in populations that bird feeding can lead to.[38] Bird therapy stations might need to tour.

There's a need for many approaches to reconnect people with nature, and it's great to be able to improve mental health simultaneously. The pathways provide a framework with many options, including for the times we can't get out into nature or for those with restricted mobility. Our research has found that a pathways-informed audio meditation can bring large and prolonged increases in nature connectedness and improve mental wellbeing too.[39]

In this instance, the focus was paranoia, the fear that others will cause something bad to happen. It is very common, but treatment methods such as cognitive behavioural therapy have limited effects on paranoia.[40] So there's a need for new approaches. A script was developed that focused on activating the pathways to nature connectedness. The narration was played over a background audio recording of a natural soundscape to help the listener imagine themselves in nature. The pathways were activated by inviting people to focus their senses. The audio meditation gently guided listeners to imagine what the natural landscape they could hear might look and feel like. As they imagined being surrounded by nature, they were invited to focus on the sensations of engaging with nature, noticing nature's beauty and being aware of the emotions evoked. People listened to the ten-minute audio meditation each day for five days. Two weeks later, paranoia was significantly lower than the control group, and a significant increase in nature connectedness was maintained, with an improvement of 17%.

If a close relationship with nature is a basic psychological need, there should be many areas where focusing on improving nature connectedness can bring benefits. I'm always looking for such opportunities, so it was interesting to contribute to work examining how the pathways could inform a programme designed to aid recovery from substance use disorder.[41] This is a condition where people can't control their use of drugs and alcohol. It affects millions of people, and various treatment approaches are used, from medication to counselling.

The Twelve Step Programme can help people maintain abstinence from alcohol and other drugs.[42] It aims to help people move away from the thinking that perpetuates addiction. The first step, accepting powerlessness, indicates the need for a 'higher power' and ultimately seeking deeper, more spiritual meaning in life. A strong sense of spirituality has been found to reduce relapse rates, suggesting that creating a spiritual connection is important.[43] However, the focus on a 'higher power' can imply ideas of a god and terms such as 'spirituality' can present challenges for some, such that agnostics and atheists have lower levels of Twelve Step Programme engagement.[44] So, alternative higher powers could prove helpful, and what better than nature? Indeed, as we considered earlier, nature is often regarded as a higher power in indigenous cultures and is used in spiritual practices. Plus, nature connectedness is associated with spirituality and wellbeing.[45] In this context, the pathways provide a framework for developing a relationship with a higher power as part of the Twelve Step Programme.

In a pilot study, the concept of a higher power of nature was used with the pathways to nature connectedness, informing the themes of each session with volunteers from an aftercare programme at a drug and alcohol treatment centre. Four weekly one-hour nature sessions were designed to activate the pathways. For example, the group considered the beauty of birds, from their colour to flight patterns. Birdsong was played, photographs displayed and feathers made available to touch. These sensory experiences provided a prompt for talking about emotions evoked by birds and the meaning they bring. The sessions improved levels of nature connectedness, wellbeing and spirituality. In comparison, those following a traditional deity approach did not show similar improvements. The approach helped those involved find peace and stillness and renewed meaning in life, which helped them connect with nature as a higher power.

This small example of the use of nature as a higher power for those recovering from addiction to drugs and alcohol provides a glimpse into a potential solution to our broken relationship with nature. This broken relationship can be seen as an addiction: addiction to consumption, dominion over nature and a lifestyle where the number of nations over-consuming resources is increasing.[46] People are also becoming more individualistic and self-centred with an excessive interest in their own needs – becoming addicted to themselves.[47] Human actions cause the warming climate and loss of biodiversity, but there is a sense of powerlessness within the forces unleashed. Accepting powerlessness and emptiness is a step to a deeper, more meaningful life than that offered by consumption – one where the sense of self dissolves and humans are part of the wider ecology. If people are empty without the rest of nature, there is a need for a new relationship with nature. There is a need to apply the pathways to nature connectedness at a societal scale.

CHAPTER 12

Scaling Up: Policies for Connection

Taking stock

The warming climate and loss of wildlife show that the human–nature relationship is broken. Our place outside nature could also play a key role in a third global crisis: mental health. For many, nature is an 'other', uncared for and sometimes tolerated, visited and sometimes enjoyed, but generally unnoticed and acting as a backdrop to modern living. Yet nature is the foundation of living – we only exist within its network.

In general, we, particularly in Britain, but also far beyond, are disconnected from nature. Earlier, I looked back to consider the history of our broken relationship – a journey from the savannahs of evolution to a life fed by nature. A difficult life, but a successful one given the 10,000 or so generations that have led to you reading this sentence now. The success of your parents, their parents and the many others who came before represents a precious chain unbroken, while billions more didn't have the opportunity to raise children. Just as every human life can be seen as precious, the same is true for the rest of nature, with millions of generations creating a thread of life within each species. Whereas this tree of life has flourished to many billions of lives for humans and their livestock, it is withering for other species as we extend our use and control of nature. In the forest of life, the human tree casts much shade.

A critical shift from generations of hunter-gatherers to generations of farmers and farm-labourers took place, changing societal structures, bringing dualistic beliefs, rational thought and increasing technology in a further revolution of science and industry. Although a challenging habitat, nature had been a home, a provider; indeed there was probably no 'nature' – it was all one natural world. The thinking of Descartes, Bacon and others can be seen to have created nature, as an 'other', to be used and controlled, or at best understood. Humans increasingly became individuals, and the new generations were of products, such as cars, televisions, computers and smartphones. Each product generation dependent on the success of its predecessor, a new survival of the fittest driving further consumption. Each product generation seeking to be noticed within a battle for attention, telling stories of how they will bring meaning to the lives of individuals addicted to satisfying themselves. People foraging in shopping centres using the same search techniques as wildlife seeking to consume in their habitat.[1] Consumer organisations recognised the importance of emotional relationships, while those looking to reconnect people with nature did not. Stripped of emotion, reduced to laws and neat equations, nature slipped from consciousness, both in human attention and its own abundance.

From that sorry story, I introduced the hope of our latent connections with nature, both seen and unseen. An invisible microbiome that keeps us well and forms part of us. The simple presence of trees and flowers that helps balance our emotions. A sensory and physical connection that merges into nature connectedness, a science focused on our relationship with nature, showing how it differs from person to person. A science that enables insight into the broken relationship, a consequence of urban living, consumption, wealth, technology and the decline of nature itself. A science that provides an understanding of the benefits of a close relationship with nature – from feeling good to playing a key part in leading a meaningful and worthwhile life, through simple moments noticing and enjoying the natural world. Those benefits extend to the rest of nature, with a close relationship with nature bringing pro-nature behaviours. The science of nature connectedness also challenges some stubborn assumptions that a close relationship with nature is built on education and knowledge. Rather, a new relationship with nature comes through tuning into and noticing nature, and activating the pathways to nature connectedness.

Addressing the global crises to achieve a good life and sustainable future is not just about restoring biodiversity and zero carbon. We had biodiversity and lost it. We had a favourable climate and are losing it. They are the symptoms of a broken relationship. A sustainable future requires a new relationship. A future where behaviours of old – the control and use of nature – are moderated, and the behaviours highlighted by the pathways to nature connectedness enhanced. This reconnection with nature needs to be based upon a fundamental recognition that we only live because of our place within the network of nature. A shift in worldview and mindset. Yet the many urban dwellers, who are often our policy-makers, are suffering from an erosion of systems thinking and have a simplistic mental model that doesn't recognise the complex interrelationships between humans and nature.[2] Similarly, this lack of thinking in terms of relationships and systems contributes to disasters in industry, health and beyond.[3]

Systems thinking has made progress in reducing harm in many domains. Yet people stubbornly focus on fixing errors rather than the deeper causes – the systemic nature of human error and the relationships between people and technology within wider society.[4] This fixation on visible problems is captured by the 'iceberg model' of systems thinking.[5] Here visible events are symptoms of unseen happenings below the surface. To prevent the symptoms of biodiversity loss and climate warming, there is a need to look much deeper to the base of the iceberg. Here, hidden deep below the surface, lie our mental models and mindsets about the way things are and how they work. These encompass our values, beliefs and assumptions, which underpin how we respond to visible events.

This model helps explain the focus on technological solutions to climate change and biodiversity loss and the inability to focus on building a new relationship with the rest of nature, despite the flawed dualistic model of the human–nature relationship being the deeper cause. Using technological development alone to rebalance the complex systems of nature is a risk, as it does not address the fundamental causes. It also continues the dominant metaphors and frames that suggest human control over nature and reinforce the flawed mindset at the base of the iceberg. The deeper issue remains, even grows, and thus the solutions are constrained, such as the opportunity to unite human and nature's wellbeing. Not in a search for transcendence, but as a new and practical approach to everyday living.

Earlier, the power of simply noticing nature was introduced. Perhaps, just as our relationship failed imperceptibly through trillions of moments, it can be renewed by creating an environment and culture of positive moments with nature. Simply noticing nature helps build nature connectedness and a deeper emotional relationship.

The butterfly affect

The butterfly effect is a term from chaos theory for a situation where large effects can develop from small changes. The classic example is the distant flapping of a butterfly altering the path of a tornado. This arose when meteorologist Edward Lorenz saw how seemingly minor changes within his weather model produced dramatic effects.[6] The butterfly *affect* thus refers to how simple moments noticing nature can influence emotional connection to nature. When noticed, simple things like a passing butterfly or singing bird affect us. We feel the effect on our emotions, which is a small step towards improved nature connectedness, mental health and pro-nature behaviours.

This dreamy world of noticing butterflies and enjoying birdsong can seem adrift from the complex lives lived by many people. The quality of these lives is perhaps yet another symptom of the broken relationship, but the evidence shows that simple moments with nature are helpful, even when working in urban environments during the winter months. Simply noticing nature can seem even further adrift from the crises of climate warming and biodiversity loss that require urgent and transformational action on a global scale. Such transformation requires fundamental political and cultural change. How can noticing a butterfly make a difference?

It makes a difference because noticing nature is the first step to repairing the broken relationship with it. Earlier, I noted how Sir Bob Watson, lead scientist of the influential IPBES global assessment report on biodiversity loss, said that humans must face a core issue. He asked: How can we become more in tune with nature? And how do we relate to nature? As we have seen, this relationship is at the heart of the climate and biodiversity crises. Noticing nature is the first step in reconnection with nature, just as the extinction of a species happens one butterfly at a time.

Many of the changes that have brought about the warming climate and loss of wildlife occurred imperceptibly over time. Each tree felled, each shovel of coal, each switch of the light. Small individual actions are both products and shapers of culture. These actions took place within a political, scientific and cultural environment that led to an exploitation of natural resources that diminished habitats and polluted the atmosphere. There will be a greater chance to limit climate chaos and the destruction of nature if a culture can be created where a passing butterfly is noticed and enjoyed by the many. Such a way of living with nature would feel good and worthwhile. People would also be more supportive of and active in the wider changes needed for a sustainable future. Butterflies can affect us in meaningful ways.

Tim Ingold explains how dualism places a fracture between 'person and organism and society and nature'.[7] Humans are both organisms within ecological systems and persons within social systems; relationships are ecological and social. Ingold suggests that social relationships are a subset of ecological relationships. Human traits and behaviours emerge from these continually evolving networks of relationships, such that, from a relational worldview, cultural variation comes from the ability to perceive and act within the environment – our ability to know and relate. So, culture is not transferred between generations but 'regrown' through training and experience of how things are done when actively engaging with our surroundings. The embodiment of an ability for relational knowing while interacting with the environment is fundamental. Yet the dominant worldview is not relational and research evidence suggests the model urban dweller's hold on ecological systems is limited.[8] This further weakens the capacities of awareness in nature that can help overcome the divide between humans and the rest of nature. Put simply, our mundane interactions with the world while doing the everyday stuff of life are essential for human–nature relationships. Connection with nature is lost when we're busy doing the other things of modern living, and in the purposeful business of objectively knowing. There's a skill of awareness in nature that has been lost.

The transformational change required for a sustainable future can be facilitated by creating a political and cultural environment that will form a new relationship with nature. A relational environment where the skill of noticing nature is continuously regrown. One where people understand the need for and benefits of reconnecting with nature. One where noticing nature matters. One where the pathways to nature connectedness can be applied at a societal scale to create a culture where nature is a valued part of everyday living. One where our relationship with nature is always considered, be it the design of urban spaces, our institutions, or our approach to transport, health and education.

From physical to emotional access

An important first step to a new relationship with nature is access to nature. At a programme and policy level, there is often a focus on this, but what exactly is access? Based on evidence that only measured time and visits, access to nature is often considered in these terms. This can be linked to the evidence that proximity of nature is good for health. Recommendations are based on what is measured, and recommendations for local green spaces are straightforward to produce and will be of some benefit. However, pocket parks and easier access to more distant natural spaces will not maximise benefits and produce the transformative change required for a sustainable future. They will do little to improve the human–nature relationship that can improve both human and nature's wellbeing.

Physical access, provided by measures such as spending time in nature, access to nature further afield and national walking trails, is often referred to as connection to nature – but as detailed throughout this book, connection is about psychological or emotional access, formally defined by the psychological construct of nature connectedness. A close and connected relationship with nature brings benefits that are often important over and above visits to and time in nature for wellbeing and pro-nature behaviours.

There are parallels to the biomedical model of health and the move to the biopsychosocial model described by Engel back in 1977.[9] In the biomedical model, people are essentially viewed as separate from the environment and affected by events – visits to nature perhaps. The biopsychosocial model recognizes the importance of psychological aspects to human health. So does thinking on access to nature move beyond the traditional biomedical approach? Just as the biopsychosocial model gives a better understanding of health, a more nuanced understanding of access can deliver better outcomes. Thinking around access to nature is starting to change, but psychological access to and one's relationship with nature is more challenging to capture in simple recommendations – such that even when there are aims and goals to form long-lasting relationships and meaningful connections with nature, the emotional aspect can be lost. It is challenging to write emotion into policy, and as considered earlier, rational thinking is preferred. When using terms such as 'connection to nature', there is a need to be clear about what is meant.

The centuries-old divide between science and emotion is still prevalent. In 2017, Michael Gove, the Secretary of State for Environment, Food and Rural Affairs in the UK at the time, set out his vision on a Green Brexit and the future of our natural environment. Three of the pathways to nature connectedness, namely beauty, emotion and meaning, were recurring themes. However, the speech implies a distinction between such themes and science: 'I grew up with an emotional attachment to natural beauty which inevitably influences my feelings towards questions on everything from architecture to ivory. But while natural beauty moves us deep in our souls, environmental policy also needs to be rooted, always and everywhere, in science.'[10]

Having evidence-based policy makes sense, but as detailed in this book, there is a science of beauty and emotion,[11] and this can and should form part of the evidence base that informs environmental policy. One problem is that much is left behind when reducing emotions to words. At its most powerful, nature produces the precious 'ooh' and 'aah' moments where we are lost for words, where we touch a life before language; we're reduced to the calls of the human animal.[12]

Earlier in the book, we saw how emotions are fundamental to human function and that regulating emotions is a continual process important for wellbeing and the choices we make. We saw how the three-circle model of emotion regulation explained how responses to flowers and trees help manage our moods. Rather than dismissing emotions, understanding the mechanisms of how nature affects us and brings mental wellbeing helps people to advocate the importance of nature and its place in policy to inform and develop effective interventions, such as green social prescribing. Being with nature is emotional, and this can and should guide policy. It should inform the types of natural spaces provided and a realisation of the need to move beyond facts, figures and science to finding joy, calm and balance in nature.

More widely, at a time of crisis in biodiversity, there's a need for narratives that show that nature matters. Given the crises in both planetary health and mental wellbeing, these narratives are essential. As an established model, the three-circle concept provides a convincing yet easily accessible story to help influence decision-makers and inform practitioners of the benefits of a close relationship with nature. Theories of natural wellbeing focused on restoration suggest short-term public health measures and pockets of green space to provide people with a dose of nature. But a culture of occasional visits to 'special' green spaces is part of the traditional relationship with nature that has failed.

The traditional thinking can also be seen at times in 2018's 'Landscapes review: National Parks and Areas of Outstanding Natural Beauty (AONBs)', published by the UK Government.[13] Encouragingly, the report states that 'The purpose to connect people to nature, and its execution, is too weak' and includes a proposal that declares, 'We need our national landscape bodies to lead the charge in connecting more people to nature'. However, the focus relates to reaching out and connecting people to nature through access and recreation. This can be seen in the ten proposals under

'Landscapes for everyone'. The mission to connect more people with nature focuses on access for all. Similarly, the proposal for 'Landscapes that cater for and improve the nation's health and wellbeing' calls for 'strong relationships with local public health teams, clinical commissioning groups and social prescribing link workers'. Promising, but again the focus is on improving access. Access is a critical first step, but there is a need to go further and consider the purpose and content of that access. As we've seen, nature connectedness brings benefits over and above visits and can be targeted for mental wellbeing.

The proposals in the report that link through to nature connectedness are quite traditional. For example, there is a proposal for 'A night under the stars in a national landscape for every child'. The purpose here is to benefit mental health and to develop pride in the natural environment and its biodiversity – if they're lucky enough to experience abundant wildlife. Uniting human and nature's wellbeing is a key outcome of nature connectedness. However, whether it's humans or nature, one night doesn't make a relationship. The evidence suggests limited success of outdoor experiences in delivering the connection that leads to mental wellbeing and pro-nature attitudes. A one-off experience is also unlikely to bring sustained benefits. But, done well, a day spent where nature thrives could provide a catalyst, and with a follow-up programme, could be used to develop a closer relationship with nature.

The proposal for children's visits includes: 'They should learn how landscapes have inspired generations of artists, poets and musicians ... a meaningful visit ... would be a chance for children to meet others from communities they may not normally meet, to learn about the nature that we all rely on.' There's a focus on learning, although as covered previously, knowledge and education struggle to deliver the connection that brings wellbeing and pro-nature behaviours. For a meaningful visit, it would be better to focus on *creating* art, poetry and music inspired by the natural landscape, through noticing nature and its beauty and telling the story of the meaning and feelings it brings. Rather than focusing on learning the history, we should help nurture and enhance the present through caring for nature. The pathways to nature connectedness provide a framework for a consistent offer and can easily be activated during these visits; indeed, many organisations are already using them to offer meaningful experiences. However, rather than a 'meaningful' visit, one should aim for a meaningful relationship with nature. Rather than 'learning about nature', bring the enjoyment and wonder of the natural world to the fore. Find stars in the everyday – in cobwebs, leaves and birdsong; in the nature children can also find everyday close to home.

Clearly, 'levelling up' access to nature is an important aim in its own right and a critical first step towards reconnection with nature, especially when it is about bringing a biodiverse natural environment close to every home. However, access is not connection. Some of those that suffer from inadequate access to nature are not necessarily disconnected from it. Some of those with good access to nature, meanwhile – often men – tend to be more disconnected. If access is conflated with connection, communities that are connected despite poor access could be unfairly regarded as disconnected. Further, the broken relationship with nature is endemic; remember that 90% of people in the UK are not entirely sure they are part of nature. Focusing on communities with poor access to nature alone is likely to miss the wider disconnected population. We can all have a love of nature.

As ever, in practice, it's not straightforward. For some, nature is a place to help manage emotions rather than develop an emotional connection. Our research group evaluated a nine-week programme delivered by Derbyshire Wildlife Trust where YMCA residents with limited access to nature got involved with nature conservation skills. Those taking part may have experienced homelessness, self-harm, exploitation, abuse, mental health problems or substance misuse. They talked of being 'away' from the stresses of complex lives and finding emotional space and calm in nature.[14] Given their circumstances, nature was a place to help manage their emotions rather than develop an emotional connection, although growing respect for nature did begin to emerge. Application of the pathways to nature connectedness must be sensitive to the circumstances.

The policy world

The difficulty we've discussed around language and thinking within policy is also evident in the UK Government's 25 Year Environment Plan (25YEP).[15] This includes a policy on 'Connecting people with the environment to improve health and wellbeing'. However, while the introduction mentions the need to 'manage our relationship with the natural world', the common language of that 'connecting people' policy revolves around access, outdoors, green spaces, time, visits and contact. The only mention of a relationship is with the 'outdoors'. Outdoors can be tarmacked and concreted, and green spaces can be sports pitches with little biodiversity. As considered earlier, wording and framing matter; they shape the way we see the world and strongly influence how we think and act. When there is a broken relationship with nature, there is a need to find ways to actively engage with nature and biodiversity – moving beyond vague terms such as 'outdoors' and 'green space', which can be seen as associated with violence, boring and 'not for us' by young adults.[16] Thankfully, there are framing nature toolkits that can help guide communication, exploring how words can help save nature.[17]

Despite this language issue, when elements of the policy have been enacted, nature connection, through the pathways to nature connectedness framework, has been integrated. For example, the 25YEP includes an environment theme for the '#iwill campaign', which aims to increase involvement in meaningful social action for all 10- to 20-year-olds. This includes working with youth and environmental sectors to help children from all backgrounds engage with nature and improve the environment. In addition, the #iwill Green Influencers Scheme aims to create lasting and meaningful connections with nature. This involves Green Mentors recruiting and supporting groups of Green Influencers within their community. These Green Influencers develop and implement environmental social action projects. Part of the training for the Green Mentors involved understanding the pathways to nature connectedness.

Similarly, in the funding guidance for the Connecting People with Nature stream of the UK Government's £40,000,000 Green Recovery Challenge Fund (GRCF), applicants were asked to work to the pathways to nature connectedness framework.[18] These funded projects, like 'Generation Green', aim to connect young people to nature and create jobs and a workforce for a green recovery. The coalition delivering the project comprises YHA (England & Wales) as host on behalf of the Outward Bound Trust, Scouts, Girlguiding, Field Studies Council and the ten English national parks, and the project makes use of the pathways to nature connectedness.[19]

Despite some of the limitations of policy language, there is increasing recognition of nature connectedness, and some important work has been coordinated by Natural England. For example, connecting people with nature was one of four strategic programmes in a Natural England policy paper on nature's recovery published in 2020.[20] This sets out to ensure that there are 'nature-rich' places close to where people live, enabling communities access to nature on their doorstep so that they can 'enjoy nature'. This approach speaks more of an everyday emotional relationship with nature. Similarly, 2020's Public Health England report into access to green spaces recognised that nature connection was more than simply spending time in nature.[21] The Government's 'Levelling Up' White Paper, with its 'pride in place' mission, provides an opportunity to use a new relationship with nature to meet the aim of improving 'people's satisfaction with their town centre and engagement in local culture and community'.[22]

There is increasing recognition of the relevance of nature connectedness for policy. This includes 2021's 'Nature and the City' by Policy Exchange.[23] Although the report retains the green space and access terminology, it recognises nature connectedness is part of access and the value of enjoying 'interacting with nature', including terms like 'emotional connection to nature'. The proposals recognise that 'access to nature is more than a single metric'. It is better understood by 'quantity' or the ease of interacting with nature measured by proximity, and 'quality' which refers to enjoyment of and emotional connection to nature. 'Quality' also includes nature-rich habitats in urban areas. This need to unite physical and emotional access can perhaps help redefine what is understood by 'access', moving from proximity to green spaces to a more encompassing opportunity to interact with urban nature. However, our relationship with nature is too important as a basic psychological need and underlying cause of the environmental crises for it to be subsumed into quality of access. Bluntly, improving access isn't fundamental to addressing climate warming and biodiversity loss; it can form a part of a new relationship with nature, but does not get to the root of the issue. Nor is quality of access a rallying cry, whereas a new relationship with nature is.

Recognition of this issue can be seen in the use of language in reports by global institutions, such as the UN.[24] Most recently in June 2022, the core importance of redefining the human–nature relationship was a key message in an independent report produced to provide a scientific basis for the Stockholm+50 international meeting convened by the United Nations General Assembly. A key message in the report is that 'Our relationship with nature needs redefining, from one of extraction to one of care. Human–nature connectedness should be strengthened in our social norms and value systems, and in how we live our everyday lives.'[25]

The report recognises that society's disconnection from nature is a root cause of the biodiversity crisis and contains a detailed consideration of redefining the relationship between humans and nature, through the science of nature connectedness. This includes the pathways research and the opportunity to target 'leverage points', which we'll consider in detail in the following chapter. Some areas for action draw on the pathways to nature connectedness to repair the broken relationship between humans and nature. This includes actions to help reconnect people, communities and societies to the nature around them. Recommendations include education policy and curricula that are informed by the pathways to nature connectedness. The breadth of change needed and policy opportunities are reflected in a suggestion for new housing developments to include opportunities for an active relationship with nature. Actions informed by these discussions are included in the summary for policymakers on how to repair and redefine our relationship with nature. The report is significant and shows the growing global recognition that the human–nature relationship is a core issue and that the science of nature connectedness can be applied to make a difference.

Bringing connection into policy

This increasing recognition of the relevance of nature connectedness also includes 2021's Dasgupta Review led by Professor Sir Partha Dasgupta.[26] It was published by HM Treasury in February 2021 and set out to assess the economic benefits of biodiversity, examine the costs and risks of biodiversity loss, and identify actions to enhance biodiversity and economic prosperity. Perhaps unexpected given the focus and source, the review recognises the essence of nature connectedness, which runs as a theme throughout.

With the focus on nature's value to human wellbeing, the review acknowledges an anthropocentric viewpoint. This is discussed with reference to alternative systems of belief and thought that go beyond an anthropocentric perspective. This discussion includes sacredness and how this can include a sense of wonder and awe that contributes to nature connectedness and human wellbeing. As a review requested by HM Treasury and focused on economics, it is commendable that it discusses comparisons of wellbeing, such as eudaimonic wellbeing and life satisfaction, and affect regulation – an often overlooked topic introduced earlier in this book. The Dasgupta Review discusses nature connectedness directly, noting the 'admirable' review of the wellbeing benefits of nature connectedness by Capaldi and colleagues. As such, the review accepts the distinction between connectedness with nature and contact with nature, noting that:

> Psychologists would appear to be on firmer ground when reporting the role contact with Nature plays in our sense of wellbeing. The influence on our wellbeing of connectedness with Nature is less assured empirically, at least as of now. The reason may be that connectedness is more difficult to achieve than making contact with the natural world. So, most studies have looked for the influence of contact on hedonic wellbeing.[27]

As outlined in this book, although starting much later than the large body of research into contact with nature, the rapid growth of nature connectedness research and evidence is already on 'firmer ground'. Solutions are emerging for increasing connectedness – with this book, perhaps the first to focus on nature connectedness directly, highlighting those and hopefully creating another island of firmer ground to build upon.

Moving on to options for change, the review again distinguishes between connectedness with nature and contact with nature, seeing contact with nature as a means to improve wellbeing and connectedness being 'an aspect of wellbeing itself'. In the section on transforming our institutions and systems, the review continues, 'Access to green spaces (they are local public goods) can also reduce socio-economic inequalities in health. Interventions to increase people's contact and connectedness with Nature would not only improve our health and wellbeing, there is a growing body of evidence to suggest that those interventions would also motivate us to make informed choices and demand change'. This captures the essential reciprocal relationship needed for a sustainable future.

Nature connectedness is also included in the review's final option for change: education. This considers emotional attachment to nature and appreciation of our place in nature – although, while the focus is on introducing children to the awe and wonder of nature, there is reference to teaching knowledge. Citing the 'teenage dip' in nature connectedness, and in language I might have used myself, the review states that 'Connecting with Nature needs to be woven throughout our lives' and there is a need to 'create an environment in which, from an early age, we are able to connect with Nature'. Finally, the final section on 'Transforming our Institutions and Systems' is notable as it often takes the form of a manifesto for connecting people with nature, with the final paragraph including the line, 'Each of these senses is enriched when we recognise that we are embedded in Nature'.

The language used in the full 610-page Dasgupta Review does not shy away from emotions and connection, and its ambition for a future where citizens can live in peace with nature is grand. From a human–nature connectedness perspective, the Dasgupta Review is a very encouraging document. However, the language of connectedness falls away in its 'Headline Messages' document. Although there is mention of interventions to enable people to connect with nature for both human and nature's wellbeing, education policy is reduced to environmental education programmes. Sadly, those engaging with the five pages of headline messages will come away with a different feel than those engaging with the much longer abridged version or full report.

It is interesting to consider the Dasgupta Review alongside the World Health Organization (WHO) publication *Nature, Biodiversity and Health: An Overview of Interconnections*, published a few months later. It includes a quote from the final paragraph of the Dasgupta Review:

> Biodiversity does not only have instrumental value, it also has existence value – even an intrinsic worth. These senses are enriched when we recognise that we are embedded in Nature. To detach Nature from economics is to imply that we consider ourselves to be external to Her. The fault is not in economics; it lies in the way we have chosen to practise it.[28]

However, this perspective does not run through the WHO publication. Although the title includes interconnections between health, biodiversity and nature, the summary reflects the anthropocentric viewpoint noted as limited in the Dasgupta Review, stating that the report 'provides an overview of the impacts of the natural environment on human health. It presents the ways nature and ecosystems can support and protect health and wellbeing'. Rather than the interconnections, the report focuses on the benefits of nature to humans. In contrast to the Dasgupta Review, there is no distinction between connection and contact, with little language around the importance of the human–nature relationship, beyond the quote from the Dasgupta Review. A linear figure on Cultural Services mentions non-material benefits such as aesthetic value and spiritual meaning, but the focus is on what nature provides for people, rather than a sense of interconnected relationships between people and the rest of the natural world. This is reflected further when considering access to nature. Unlike the Dasgupta Review, the WHO report contains little on interconnectedness and the opportunity to build both access and connection to move beyond the one-way benefits nature provides to human health. The conclusions of the WHO report do mention more reciprocal relationships, such as 'sustainable behaviours that benefit nature and health', 'simultaneously providing human wellbeing and bio-diversity benefits' and 'promoting benefits for both human health and the natural environment', but the unity of human and nature's wellbeing could be a stronger theme throughout – especially as other global institutions are heading in this direction.

Although it is a review of the economics of biodiversity, the Dasgupta Review sets out, understands and embraces the emotional and soulful relationship between people and the rest of nature and how that is key for human and nature's wellbeing. In contrast, the WHO report on the interconnections between nature, biodiversity and health sets out the importance of nature for health and thereby the need to protect it. To use the terminology of Arne Næss, the Norwegian philosopher, this reflects a 'shallow ecology' that considers nature important because nature is useful to humans. In comparison, 'deep ecology' sees the intrinsic value of nature.[29] Yet the Dasgupta Review recognises a 'reverential ecology' where nature is sacred, and all things are related, and we, as humans, feel at home and at ease in nature.[30]

As found in the brief 'Headline Messages' document of the Dasgupta Review, the challenge lies in retaining and reflecting those essential deeper elements in policy recommendations. There is still a need for wider recognition, understanding and acceptance of the human relationship with the rest of nature. In practice, we separate humans from nature and reinforce the current worldview. A relational worldview moves from entities towards embodied experiences, processes and relationships. A new outlook requires a new language, a language of nature connectedness compatible with policy proposals, whether that comes from policymakers finding a language of connection and emotion or researchers writing in policy terms – or ideally, working together. New language and thinking can foster more holistic and dynamic visions of a closer relationship with the rest of nature that can inform decision-making for a more sustainable future.[31]

It is possible to build a language of nature connection into policy. In 2021, I worked with the Mental Health Foundation on the nature-themed Mental Health Awareness Week. As well as a research report,[32] a nature and mental health policy briefing was produced.[33] The Mental Health Foundation's policy team did an excellent job of turning the research into policy proposals. These were interesting and enjoyable conversations. In an associated survey, the Mental Health Foundation found a clear public appetite for change, with three-quarters of people thinking that the Government should encourage people to do more to connect with nature. The briefing noted our research showing that a meaningful connection with nature is often more important than time spent in and visits to nature. Therefore, the briefing makes a case for 'prioritising connection with nature as the main goal for our nature and mental health policies'. The recommendations for how government policy can facilitate greater nature connectedness fall into five areas.

Firstly, we have facilitating connection with nature through moving beyond access to nature and towards engaging with nature, such that nature connection is the core principle that drives policies relating to mental health and nature. Second is protecting the natural environment and restoring biodiversity through ambitious targets to halt the decline of species and habitats, with the delivery of biodiversity gain prioritising deprived areas to bring the wellbeing benefits of nature to the communities that need it most. The third is improving access to nature, which includes improving safety and the quality of green spaces and parks, so that access to nature will be guaranteed for the broadest range of people. Fourth is using urban design to improve the availability and visibility of nature through facilitating an everyday connection with nature, especially in urban areas. Fifth is building a life-long relationship with nature through nature being a part of the learning process and school grounds providing access to nature, especially in secondary schools.

Working with the Mental Health Foundation and their briefing helped me to refine and simplify my thinking; after all, I am a researcher, not a policy expert. This was aided by writing guest blogs with short word limits and a policy focus. Building on the Mental Health Foundation's mental-health-focused briefing, I arrived at five broad implications for policy that can help bring about a new relationship with nature:

- Foster a culture of connecting with nature across the public realm – from an arts policy that celebrates nature, to nature connectedness being a standard metric for wellbeing.
- Use biodiversity to unite the wellbeing of people and nature by bringing nature recovery networks into urban areas for an abundance of wildlife to notice.
- Improve access for connection – moving beyond simple access to creating 'habitats for connection', providing meaningful engagement with nature on everyone's doorstep.
- Design urban spaces around the pathways to nature connectedness to prompt people to notice, engage with and care for nature.
- Build a life-long relationship with nature – addressing the 'teenage dip' by bringing nature into secondary schools and ensuring everyone understands the value of nature for keeping well.[34]

CHAPTER 13

Tools for Change

Leverage points

There has been some promising progress in the UK and there are signs that key institutions are starting to understand that our relationship with nature is the missing link, a single and tangible target within a multitude of symptoms. The climate and biodiversity crises cannot be solved without addressing the causal issue: how people relate to the wider natural world. However, as we saw earlier in the book, our fractured relationship with nature has a long history. Many are yet to recognise the importance of this relationship, continuing with a focus on people or nature. Furthermore, although tangible, we do not know what a future with a close relationship with nature looks like. Living in harmony with nature is alien to many and difficult to envisage in our modern world. This work to imagine such a world is beginning. It is recognised globally by the UN[1] and IPBES, the intergovernmental body established by the UN, is undertaking a transformative change assessment, which will consider solutions for the underlying causes of biodiversity loss, such as the human–nature relationship.[2]

The severe consequences of the ingrained breakdown of the human–nature relationship require a paradigm shift and societal change.[3] Thus, bringing about reconnection with nature requires approaches that can effect significant changes at scale to create nature-connected societies. The pathways to nature connectedness have informed successful large-scale campaigns and visitor experience programmes, but there is a need to scale up beyond these public engagement programmes. There is a need to explore the relevance of the pathways at a societal scale and apply systems thinking to stand a chance of achieving societal change.

Clearly, societal change is a mighty ambition where effort needs to be targeted in areas that might bring the maximum benefits. When something large needs to be shifted, it is helpful to use a lever. So, when taking a systems approach, it is important to consider the places in the system where maximum impact can be gained. This can be done by thinking about 'leverage points', where small changes within a complex system (e.g. an economy, a city, an ecosystem) can bring about large effects. Donella Meadows was regarded as an influential conservation and environmental thinker, and in a widely cited article from 1999, she described 12 leverage points. These ranged from 'shallow' places where interventions are relatively easy to apply but less impactful, to 'deep' places where interventions are difficult to apply but have the potential to deliver transformational change.[4] Leverage points also help work across disciplinary boundaries by providing a shared vocabulary and narrative.[5] The order of the 12 leverage points identified by Donella Meadows is 'slithery', but they can be considered in four broad groups, which make things a little easier to think about.[6] These concepts can be challenging to understand, though – creating a nature-connected society won't be easy. Yet straightforward visions of the future will be required.

The shallowest leverage points, which are easy to utilise but have less potential for impact, are 'system parameters'. These include the standards and flows that govern how a system works; some examples will be helpful. Air quality standards are one example of a standard, while a flow could include the flow of nitrogen fertiliser into the soil and subsequent run-off into rivers. Through setting parameters, governments can adjust the quality of the air we breathe or control fertiliser run-off. The amount of land set aside for nature conservation is another example of a parameter. These adjustments could lead to more rewilding and cleaner air, but the old system, with the same goals, would still exist. Generally, changing standards and parameters does not create transforma-tional change. However, certain system parameters can make a significant difference when they interact with deeper leverage points. For Meadows, system parameters such as standards were 'dead last' on her list of powerful interventions. They simply fiddle about with details, akin to rearranging the deckchairs on the *Titanic*, and rarely change behaviour. If a system, such as the ecological system, is out of control, standards don't stabilise or put a brake on it. Dirty air becomes less dirty. Rewilded land increases, but consumption and dominion over nature continue. Yet 99% of our attention and energy goes to the debate and design of parameters.[7]

The third most powerful group of leverage points are 'feedback loops'. These are the interactions between the various elements within a system. Negative feedback loops are ubiquitous, both evolving within the natural world and created by humans in order to control systems. A thermostat that controls your home's heating is a common example. As such, feedback loops have a goal (temperature), a monitoring device (the thermostats) and a response mechanism (often, for the time being, a gas boiler). Human encroachment into the habitats of endangered species is used as an example of a negative feedback loop by Meadows. Whereas negative feedback loops are self-correcting, positive feedback loops are self-reinforcing, sources of growth and collapse; the more red kites that nest successfully, the more red kites there are to nest. The response of the climate to levels of emissions is another feedback loop.[8] Interventions that target feedback loops include increasing the numbers of natural predators to control crop pests, taxes on pollution, and monitoring and reporting environmental damage. As mentioned earlier, there are reinforcing feedback loops related to nature connectedness and noticing nature. People who have a closer relationship with nature also have better attention capacity and explore natural views more, thereby noticing nature more and building their relationship further.[9] Similarly, earlier we saw how actions to create visible biodiversity boost noticing nature, thereby bringing people closer to nature, which motivates further actions for biodiversity.

The second most powerful group of leverage points are 'design characteristics'. These include access to information, rules, constraints, incentive schemes, and the power to change and self-organise; in biological systems, this is called evolution. This can include the social structures that govern information flows, manage feedback and system parameters. Examples include visible smart electricity meters that provide access to information, delivering a new feedback loop. Or giving easy access to information such as school league tables. Compelling feedback can change behaviours. This can come in the form of public access to knowledge about harms to nature, such as the source of products or information on the policies and regulations governing the use of natural resources. Knowledge is power and these aspects of system design relate to human cultures and social evolution, potentially towards a more nature-connected society. The loss of wildlife can be seen as the public having had a lack of information, lack of feedback on the state of nature and lack of freedom to respond. Knowledge and information can support the restoration of nature – for example, through the ability of communities to manage the use of local land or to manage the shallower leverage points, such as creating their own feedback loops or local rules. This freedom to self-organise is very powerful, indeed essential for long-term survival, but can be seen as a loss of control by authorities. Like rewilding, the value comes through giving up control, which requires incentive schemes such as the Local Nature Recovery scheme of the Agriculture Act 2020, providing farmers with incentives to restore wildlife habitats.[10] Changes in governance structures that allow more people to participate allow a wider contribution to nature conservation, but also change values towards nature.[11]

Finally, the most powerful and deepest group of leverage points are 'intentions' – the underpinning worldviews, values and goals embedded in the system as norms that shape the emergent direction of that system. So, although not necessarily an explicit goal of the socio-economic system, economic growth can be the emergent intent if the system, its worldviews and purposeful behaviours support that as its main path. Given the subsidies paid to farmers for farmable land, an emergent property of the European Union's Common Agricultural Policy can be seen as the removal of pockets of nature on farmland.[12] Intentions and goals define the shallower leverage points too. There is also a hierarchy of goals, some of which can compete – for example, the goals of a competitive market can trump the goals of individual corporations or populations.

When contemplating our broken relationship with nature earlier, we considered the evolution of worldviews, from hunter-gatherers to the Scientific Revolution and beyond. Meadows highlights how deep-set beliefs about the world arrive from the often unstated yet 'shared idea in the minds of society':

> Nature is a stock of resources to be converted to human purposes. ... One can 'own' land. Those are just a few of the paradigmatic assumptions of our current culture, all of which have utterly dumbfounded other cultures, who thought them not the least bit obvious.[13]

Interventions at the deepest leverage point totally transform systems. But paradigms are the hardest aspect of a system to change. Still, that change can come swiftly, from a single individual providing a new way of seeing – in science, for example, with a Copernicus or Einstein. Societies resist a paradigm shift. But although there is certainly no silver bullet to turn around self-destructive trends, paradigm shifts that once seemed unthinkable do occur.[14] Through repeatedly focusing on the failings of the old paradigm and through people in visible and influential positions speaking assuredly from the new paradigm, change can happen.[15] Hence, through feeding back information on the broken relationship with nature and through powerful advocates providing a vision for a new relationship with nature, we can effect a transformation. Yet, as noted, it can be difficult to envisage what a close relationship with nature within a modern technological society might look like.

One example of a paradigm shift that has encountered resistance is rewilding. From a leverage points perspective, rewilding projects involve a paradigm shift in our relationship with nature, not simply different land management standards. In *Wilding*, Isabella Tree sets out the clear intentions and change in worldview that led to the pioneering rewilding of the 3,500-acre Knepp estate in West Sussex.[16] Taking a leverage points perspective, this involved the removal of constraints, giving up control of the landscape and letting nature take over or self-organise. To create new habitats for wildlife, free-roaming ponies, cattle, pigs, fallow deer and red deer were introduced as proxies of the large mammals that once roamed the landscape. This paradigm shift brought significant local resistance and anger. The land was being abandoned to 'weeds'. The aesthetic character of the landscape was destroyed. 'Beauty' was lost. When local villagers were surveyed, most considered themselves to be nature lovers. But it was love on their terms. In addition to creating flashpoints with local people, the rewilding created self-reinforcing positive feedback loops with increased wildlife and habitats, bringing yet more diversity. The project's success provides access to information, new parameters and standards for other rewilding projects, such as the introduction of free-roaming grazing animals. Isabella Tree and her husband Charlie Burrell now speak assuredly about the new paradigm, bringing to life a vision for a more sustainable future.

It is interesting to consider what will happen when rewilding projects are brought to where people live, a sensible vision for a more sustainable future. Here a new paradigm abuts the current paradigm where green spaces provide a resource for cycling, running, dog walking and the like[17] – activities that people enjoy but which disturb nature. Proposals for the future are expressed and considered based on how natural landscapes are currently used, without experiencing the vibrant spectacle of a rewilded landscape such as Knepp. Attitudes and perceptions of rewilded landscapes change once they are experienced. Many of those living close to Knepp have become supportive as the birdsong embraces them, with those who once felt the landscape had been wrecked now finding beauty.[18] In designing schemes based on the current paradigm, there's a risk that urban wilding, while being of some benefit to biodiversity, will remain the mere backdrop of human recreation rather than being transformational for biodiversity and connectedness.

Although a couple of decades old, the leverage points approach is still used. It has been suggested that sustainability research and policy has focused on treating symptoms and mainly addressed the shallower leverage points rather than root causes.[19] The current paradigm and worldview can lead to the symptoms being treated with new technological solutions that allow further control.[20] As we have seen, human–nature connectedness is the root cause and is considered a leverage point for sustainable transformation, being the topic of a special issue of the journal *Ecosystems & People*.[21] Although yet to make an impact within widespread dualistic thinking, the human–nature relationship and people's connections to nature are seen as one of three key realms for transformational sustainability interventions, alongside restructuring institutions and rethinking how knowledge is created and used.[22]

The power pathways

Transformational societal change for a close relationship with nature is a challenge of the highest order. In addition to leverage points, there is a need to establish which of the five pathways to a closer relationship could have the greatest impact and societal relevance. To do this, we considered the relevance of each pathway at the individual and societal level and their potential to be applied at deep leverage points.[23] Let's first look at the large-scale relevance of each of the pathways.

As care and compassion for nature are the overall aim of a new relationship, it might be tempting to think that the compassion pathway provides the best opportunities for deep leverage and societal change. However, compassion and care for nature are more of an end point; those engaged with pro-nature conservation behaviours typically have higher levels of nature connectedness. So before attempting to engage people with pro-nature conservation activity, there is a need to increase connectedness. That said, compassion and care for nature go beyond pro-nature actions. Humans are a social species. Social connections are important to us and formed through similarity and emotional bonds that build our capacity for cooperation and compassionate helping – overcoming the destructive tendencies of the human species.[24] Therefore, focusing on the similarity of people with nature is worthwhile. For example, research shows that anthropomorphism

– that is, giving nature human qualities – can drive nature connectedness and pro-nature action.[25] A culture that accentuates the similarity of people and the rest of nature together with the provision of everyday opportunities for people to care for wildlife would be beneficial.

If care and compassion for nature is the final goal, then contact and sensory engagement with nature can be seen as the first step along the pathways to connectedness. As noted earlier, readily accessible nature does not demand to be engaged with, and it has been suggested that contact with nature provides shallower leverage.[26] So, while sensory contact with nature can have a large scale of relevance through improving access to nature, engagement needs to be fostered to deliver societal impact on the human–nature relationship. As seen earlier, people don't tend to notice nature, but it can deliver sustained improvements when they do. Fostering engagement requires the building of intrinsic motivation, targeting deeper leverage points and activating other pathways to nature connectedness. This highlights that the pathways to nature connectedness rarely work alone. Sensory contact is a power pathway because it's the first step to eliciting emotions, bringing meaning, compassion and encouraging one to notice nature's beauty.

Beauty emerges as a strong theme when people are asked to notice 'good things in nature'. However, its impact comes through working with other pathways, such as when deriving meaning or evoking emotions. Engagement with beauty is emotionally based.[27] When measuring engagement with nature's beauty, the conscious and physiological emotional response is rated – for example, feeling the heart racing.[28] It is therefore likely that the beauty pathway provides a relatively shallow leverage point. Instead, beauty needs to be available for sensory contact to provide meaning and evoke emotions.

The emotion pathway is central to the human relationship with nature, and emotions are fundamental features of human function, targeted at scale in advertising to sell consumer products. It is clearly a power pathway. Many readers will remember the 'drumming gorilla' advert for chocolate in 2007. The abstract nature of the ad meant it sat unused for nine months until it was shown to evoke emotions in high-tech testing. It was then released to great success and sales soared.[29] Targeting people's emotions to give consumer products meaning has already brought societal scale change through developing consumerism and self-absorption in Western society. Understanding hedonic experiences and the pursuit of feelings and fun is a key focus of consumer marketing.[30] Yet many do not believe the emotional relationship with nature matters. That same focus could highlight the essential meaning of our emotional relationship with nature. The benefits of a worthwhile life and sustainable future shouldn't need a hard sell, but there is a battle for attention, an addiction to our consumerist selves and a tired assumption that knowledge changes behaviour. The shift from the desire for consumer goods to experiences provides an opportunity to promote pathways focused around experiences in nature, ahead of ones based on the control and use of nature.[31]

Finally, meaning is another pathway to nature connectedness. Humans constantly seek to understand and make meaning of the world around them and the relationships between things. Indeed, forming a mental representation of the relationships between things is a reasonable definition; meaning connects things.[32] Often, meaning is seen as central to human life.[33] Together with sensory contact and emotion, meaning is a power pathway.

Meaning relates to cultural aspects of our lives. Earlier in the book the decline in nature words in cultural products, such as books, films and songs, was noted as a sign that nature means less and less in people's lives. Meaning is part of deeper connections with nature, with aspects such as folklore and engagement with the signs and cycles of nature. These cultural conceptualisations of nature, when widespread and celebrated in societies, provide powerful metaphors and impact on individual connection.[34] Meaning can have great resonance at a societal scale, but it is difficult to change the meaning of nature in society. There are ways – for example, cultural programmes that focus on developing these deeper relationships with nature, fostering the celebration of those relationships, and encouraging an increased appearance of nature in cultural products and societal systems. This would involve changing the values and goals of organisations such as those that fund the arts and deliver healthcare.

Lessons from the consumer world

Before we move on to consider how the power pathways might be applied at the various leverage points to create a nature-connected society, let's consider a couple of wider points. Throughout the book, from the broken relationship with nature to the power pathways, stories from the consumer world have appeared. We've heard how the carefully crafted 'emotional architecture' of social media is used to engage and capture attention. We looked back to the origins of the targeting of emotions to mould public desire, creating a dominant consumer culture across society. Experts in consumer behaviour also understand that the social dimension of emotions is an essential factor in consumption.[35] When identifying the power pathways, we've seen how this targeting of emotions to give consumer products meaning remains a key focus of consumer marketing. Emotions aren't the only route to creating meaning. Earlier in the book, we also considered the animalistic worldview, where the non-human world has life and spirit. Once again, while some discuss the 'error' in this belief, marketing exploits it, finding that animism drives positive brand responses, is easy to activate through simple cues and that it 'breathes life into brands and products' and facilitates attachment.[36] So, it will come as little surprise that an influential framework of factors in consumer behaviour from 1982 includes several keywords from the pathways to nature connectedness – namely sensation, emotions, beauty and meaning.[37]

It is an uncomfortable truth that researchers in consumer behaviour have pieced together the essential elements of psychology more effectively than those interested in the human–nature relationship. The pathways to a consumer–brand relationship are much the same as those to the human–nature relationship. Nature can even be used to increase purchasing and prices.[38] It is an obvious point, but consumer marketing understands people better than nature conservations and environmentalists; but psychologists too have paid too little attention to our relationship with the natural world. The failing relationship with nature is pervasive. Taken together, this collective failing explains a great deal about the human-created crises the natural world now faces. The pathways to nature connectedness are decades behind pathways to consumerism. Change is needed.

Social movements

One way change can happen is 'bottom-up', through social movements – events that create new feedback loops, and challenge social systems and worldviews. Social movements also activate the power pathways, such as emotion and meaning. From a cultural perspective, meaning-making refers to collective interpretation that allows people to understand. Meaning includes our thinking about what is true and false, moral understandings of right and wrong, and understanding what is beautiful or ugly. And also, what is to be human alone or as part of nature. We find and impose meaning. Meaning is a source of controversy and a spur to action. Meaning-making is an important part of social movements that 'bubble up' and create change.[39] Social movements actively make meaning and challenge established meanings.

There has been a rapid rise in movements that create social change over the last decade.[40] So why have nature lovers, the millions who join nature conservation organisations, failed to bring about the social change that might halt the decline of nature? Perhaps it is partly because it is seen as a 'nature problem' rather than a human–nature relationship problem that requires social change. Either way, presuming those millions do care about the state of nature, an essential element for social change is present: networks with many individuals and organisations with a range of skills and perspectives but a common cause.[41] A cause that has had years to define its narrative and demands. Some movements, on the other hand, can now grow too fast through social media. Bringing in an influx of new activists and voices before the narrative and demands are well established makes sustained change difficult.[42]

Those that study such things suggest that social movements work when they dominate attention, are emotional and have a meaningful message. A shared focus on grabbing attention brings the movement solidarity and emotional energy, thereby becoming symbolic and meaningful.[43] Surely there are enough nature lovers to achieve this. Hundreds of thousands of people are happy to count birds in the RSPB's Big Garden Bird Watch or enjoy themselves in nature through the Wildlife Trusts' 30 Days Wild. Yet ultimately most remain bystanders to nature's decline.

Older work on social movements suggests that when there's a critical issue (Stage 1) that leads to known failures, upsetting events, campaigns and growing public recognition, it only builds stress in the system and creates the conditions for change (Stages 2 & 3). For change there is a need for a trigger event (Stage 4) that starkly makes the public aware of the problem, such as the siting of cruise missiles in the UK in the early 1980s, the Chernobyl disaster in 1986 or the killing of George Floyd in 2020. With growing public opposition to the current state, the issue makes it onto society's agenda, and a social movement suddenly takes off. Yet change does not happen (Stage 5); that requires broader public support for the alternatives which the movement can exploit to place the issue on the political agenda (Stage 6). With consensus forming and the majority wanting change, those in power make some reforms (Stage 7), but real social change may require further trigger events, and the struggle continues (Stage 8).[44] The regular extinction of a species must simply be upsetting to some, and in a humanist society, the state of nature is perhaps not a social problem.

The environmental movement Extinction Rebellion 'uses non-violent civil disobedience in an attempt to halt mass extinction and minimise the risk of social collapse' through political change.[45] It appears they are creating stress in the system, needing and perhaps creating trigger events to raise public awareness through dominating attention and emotive messages. It could be that they hope the direct action leapfrogs to the political reforms of Stage 7. However, do they have a unifying story that engages sufficient public support for the proposed changes? The concept of a new relationship with nature is unifying in essence, can target the emotions that spark action and provides a story to spread and share.[46]

Whether in terms of social movements, consumer products or the natural world, humans are deeply affected by emotions and stories with meaning. We want to believe our lives are worthwhile and meaningful, either through the stuff we own, things we do, movements we're part of or hopefully through our relationship with nature. The power of emotions and shared stories has been used to bring millions of people together, create consumer culture and disconnect us from nature. However, our hidden story is that we are an integral part of nature, and that connection with it is crucial. For a sustainable future with a closer relationship with nature, societal values, social structures, feedback and policy need to tell that story. How can the power pathways be applied at leverage points to make a nature-connected society real?

CHAPTER 14

Creating a Nature-Connected Society

One of the problems with the increase in calls for people to 'reconnect with nature' is there is often fragmentation or little definition around what nature connection is. Even terms such as 'connection to nature' place people outside nature. Here our language constrains solutions; we don't have words such as *Ren*, the Song dynasty term for living in harmony with the natural world.[1] A new language of nature connectedness would be helpful, with powerful metaphors that capture the reality of our place within nature. That said, connection to or with nature is becoming a widely used phrase. I'm yet to hear a convincing alternative that comes without baggage or a transcendental feel that is a poor fit with the reality of everyday urban living. However, the calls for reconnection are often vague, with little guidance towards achieving nature-connected societies for a sustainable future. My suggestion is, helped by the evidence presented in this book, that the psychological construct of nature connectedness helps define and understand people's connection with nature, providing a measurable target within this fragmentation and approaches that can provide sustained benefits to both human and nature's wellbeing.

Having noted the lack of detailed guidance, it is perhaps recumbent on me to propose some ideas for improving nature connectedness at a societal scale. As this takes me outside my expertise, I find these proposals are best done in conversation with other experts, so the ideas presented in this chapter are conversation starters rather than a refined manifesto. I am also aware that there are many other approaches to change beyond leverage points. There can be valuable input from many perspectives. Those conversations between researchers, policymakers, communities and many others should imagine visions of a future with a close relationship with nature and then create opportunities for action and active engagement.

Power pathways at leverage points

That we are all part of the relationships of the natural world rather than humans in competition is a compelling narrative. Having separated ourselves from nature, exploited it and sliced it into component parts to examine it, the next chapter in the human story should be about reunification.[2] A new enlightenment of integration and a return to an animism based on our scientific understanding that we are all part of something, and many things are part of us. This new meaning of life provides a new paradigm and is a powerful pathway to change that sits naturally with the deepest leverage point: the underpinning worldviews, values and goals embedded in our society as norms. Realising we are part of nature and living that reality through seeking a closer relationship with it would lead to change.

Society comprises many complex systems such as institutions, organisations and corporations, each with their own values and goals. One by one, people running various organisations would need to see the failings of the old paradigm and advocate the advantages of a closer relationship with nature. This may seem highly improbable, but organisations are aiming for a zero-carbon future one by one. The goals of an organisation are simple to write but most difficult to change. Think about education. Modern schooling came about during the Industrial Revolution, to prepare children for work.[3] What should the children in our society learn at a time of climate warming and biodiversity loss? Compare those thoughts with the priorities of the UK Government's Department for Education (January 2022):

> *Our priorities:*
>
> Drive economic growth.

> Level up education standards.
>
> Support the most disadvantaged and vulnerable children and young people.
>
> Provide the best start in life through high-quality early education and childcare.[4]

These priorities are headed by the existing paradigm, a worldview that feeds many factors in the broken relationship with nature, detailed previously. Clearly, the economy is important, yet there is an urgent need for a sustainable economy. Analysis of recent government proposals for sustainability has found a foregrounding of economic concerns with an over-reliance on science, knowledge and skills.[5] Standards and equality are also important, but at best, a sick planet will likely lead to a poorer world, with greater inequality and loss of opportunity. At worst, during a full-blown natural crisis, standards and 'levelling up' might be seen as nicer problems to have. Surely, with the natural world in crisis, a priority of education should be how to live in greater harmony with nature. Revising these priorities to include a sustainable relationship with nature would be a simple change of wording that would help create a new paradigm. As a deep leverage point, that change of goal and values would feed through to curriculum design and teaching. But such a simple change would be very difficult to achieve, although the recent notion of 'levelling up' was soon added as a priority.

Corporations will include zero-carbon goals in their priorities as it becomes a required standard and social norm. However, as noted earlier, biodiversity loss gets far less attention, so the state of the natural world and the worldview underlying the current human–nature relationship is unlikely to change. Yet a sustainable future also needs biodiversity, and one underlying cause of the loss of wildlife is the human–nature relationship. Therefore, organisations could aim for 'sustainability plus', where zero carbon is combined with 'max nature' through a closer relationship with the natural world. Indeed, restructuring institutions to favour sustainable behaviours has been identified together with reconnecting with nature as a key deep leverage point for a sustainable future.[6] Finally, while meeting their environmental commitments, focusing on a closer relationship with nature will also be good for employee wellbeing. Such corporate social responsibility has also been linked to improved financial performance through improved nature connectedness.[7]

As a basic psychological need, nature connectedness is well positioned to inform the values and goals of our systems at a time of crisis in the natural world. This can include models of health that unite human and nature's wellbeing. Embedded as a goal and worldview within organisations, delivering health and wellbeing would lead to the inclusion of the enjoyment of nature in health and social care delivery, which also increases the meaning of nature in society more widely.

Perhaps some people and organisations would see rewriting their goals as a simple yet meaningless 'lip service' that will in reality change little. The leverage points approach can help persuade people of the importance of such changes and the need for the goal to filter down through organisational design, information flows and operating standards. From an institutional perspective, as a measurable construct, nature connectedness can be a key performance indicator. Targets can be set, and the pathways used as a structured means to inform strategic plans, design, feedback and operating standards.

One trend that could help this ambition is the change in strategic thinking within organisations over recent decades – from a competitive perspective with an internal locus, to a connected and then holistic perspective with a more relational external focus. Such a perspective directly impacts the organisations' strategic processes. A fundamental sense of connection with nature can impact strategic thinking, permeate organisations and integrate with the core purpose and operational processes.[8]

Values and goals inform the second most powerful group of leverage points: the design of systems and organisations. The way our institutions and social structures are designed has a powerful influence on social norms, including our relationship with nature. The power pathways of sensory contact, emotion and meaning can be integrated into social structures through the design of information flows, rules, constraints, incentive schemes and community programmes.

Community-based approaches are important. People tend to be happier when they see value in the local neighbourhood. Communities can foster a sense of belonging, bring meaning, provide purpose, and unite human and nature's wellbeing through protecting their local ecosystem.[9] This leverage point includes giving communities power to change and self-organise. There is a growing trend of communities reclaiming urban space and creating 'urban commons'. These allow citizens to manage their local spaces and reframe urban living through creating socially just spaces that provide a more sustainable way of living.[10]

Through the large institutions we interact with, many aspects of our daily lives could be designed to help create a new relationship with nature by activating the power pathways. The pathways could inform education curricula. It is encouraging that this and other suggestions in our research were picked up by the Stockholm+50 evidence review.[11] Education could take children out into wilder school grounds, start plant biology with a sensory experience, consider the physiology of how nature helps manage our emotions, explore how the meaning of nature has been represented in art, music, poetry and folklore. It could explore the history of the human–nature relationship and its exclusion from our models of health and wellbeing within a worldview of interrelationships rather than the dualistic 'nature and us'. Such a change could fulfil existing education priorities, giving children the best start in life, supporting the disadvantaged, levelling up standards and creating a generation with a new perspective on a sustainable and circular economy.

One-health models that unite human and nature's wellbeing could drive the design of health and social care services. These services should recognise the role of nature in managing emotions and keeping people well, and utilise the power of sensory contact and positive emotions for mental health through green prescriptions. The power pathways and nature connectedness as a basic psychological need could also inform new approaches to care, such as those we discussed earlier for substance use disorder. As nature connectedness is related to happiness, feeling that life is worthwhile and lower levels of significant sources of ill-being such as depression and anxiety, an individual's relationship with nature could be a valuable tool when assessing people's wellbeing.

Recognising the links between cultural expression and the pathways to nature connectedness, incentives such as funding for cultural programmes could use public spaces to celebrate nature. This could involve re-imagining traditional festivals, for example wassailing in the UK, and include the traditions of wider communities. In the UK, the National Trust is inviting people to emulate Japan's Hanami, the tradition of celebrating blossom. Through #BlossomWatch the Trust hopes to make it an annual tradition, asking people to share the joy blossom brings and enable everyone to celebrate nature together.[12]

Earlier, we saw how land use plays a key role in our emotional connections to nature, that teenagers see natural spaces as 'not for them', and that there is a need to consider the way we manage and use land. The ability of communities to self-organise and manage the use of local land for their relationships with nature provides a powerful leverage point. However, less than 1% of the population owns half of all England.[13] This prevents local management, with rights to roam and enjoy nature being limited too. Telling the story of the decline of wildlife in England in *Our Place*, Mark Cocker suggests that the villains are the moneyed landowners, industrial agri-businesses and politicians who speak for them.[14] Landowners play an essential role in our relationship with nature. A reconnection with nature requires a revolution in how we access and manage our land for recreation, farming and nature conservation. By mapping how close people live to nature reserves, it has been found that they are an important factor in nature connectedness.[15] So enabling communities to create, manage and celebrate them would be a useful step towards a new relationship with nature.

Earlier in the book, we considered the relationship between cultivated land and nature connectedness across Europe. Several countries with higher levels of nature connectedness had different approaches to farming – such as Bulgaria, with a history of state-run collective farming; France, where cooperative agriculture is more common; and Spain, where the regional Madrid government introduced participatory agricultural laboratories as a public initiative to restore the relationship between rural, peri-urban and urban areas and the natural environment through agrarian activities. Through increasing the social importance of the agricultural landscape and providing meaningful interaction with nature in people's daily lives, these initiatives have been linked to higher levels of nature connectedness.[16] This provides an example of how incentive schemes could help deliver a new relationship with nature.

In addition to incentive schemes, access to information and knowledge is an important leverage point. The importance of information can be seen in how it is sometimes hidden or discredited. For example, in 'Climategate' in 2009, leaked emails were leapt on to discredit data, affecting public belief and delaying measures that may have slowed climate warming.[17] The role of information and need to design in new feedback loops about the human–nature relationship should be considered, giving the public access to data on the state of nature. It is interesting that in *Our Place*, Mark Cocker highlights the need to make nature and wildlife publicly accessible rather than hidden 'behind a Latin wall'. As well as 'what' nature there is, local people should be aware of 'how much' nature there is, with information on the places 'where' nature can be seen. This can be combined with the power pathways through more meaningful maps showing the positive emotions evoked by nature rather than simply the route from A to B. Information can create meaning and stories, making hidden nature visible and enabling feedback on the positive links between people and local wildlife. Sensory technology can give local people access to the families in nest boxes, the parents' visits and when the offspring fledge. Providing this information in public places, at bus stops and via smartphones could play a role.

Designing in the information for feedback loops brings us to the third most powerful group of leverage points. One key reinforcing feedback loop concerning the broken relationship with nature is the 'extinction of experience'. Shortening feedback of information on positive interactions with nature, particularly information related to sensory contact and emotions, can counter the extinction of experience. A small levy on consumer advertising could recruit creative minds to create positive feedback through ad agencies for nature – groups creating and engaging people in new narratives that highlight the meaning of nature to humans. Just as the reduction in the experience of nature has permeated through culture and society, reinforcing a social norm, positive feedback loops could be used to build a closer relationship with nature. Design that increases engagement with nature will be reflected in cultural feedback related to the meaning pathway, building the orientation to engage with the natural world. This can be done while disrupting feedback loops related to the use and control of nature. As Laurence Rose puts it, we must 'stop normalising loss' and fight the 'cultural indifference' to the loss of wildlife. All extinction starts with local loss, but we get little feedback of such losses.[18]

The final and shallowest leverage points are the standards, policy and infrastructure that govern how organisations and societies work but alone don't change the current paradigm a great deal. Although standards provide weak leverage, they can offer many opportunities to apply the power pathways – from guidance for their use in funded nature engagement projects, to informing outdoor activity at school, and accreditation of parks and green spaces for the opportunities they offer for nature connection. Standards can also be creative; rather than a vague green space within a kilometre, why not aim for everyone to share their neigh-bourhood with at least 60 bird species? That motivates a richer natural environment and provides opportunities for community monitoring and new feedback loops. There's also the potential to highlight the negative relationships with nature through standards, such as product labelling that includes ingredients or highlights practices involving the use and control of nature. Such examples might make a difference, but they will not lead to the paradigm shift and societal change required for a more sustainable future.

Infrastructure is slow to change; however, policy advocating a new approach can help show what is valued and also helps direct funding. Policy change may be relatively ineffective in influencing the human–nature relationship. Still, it can send a clear message, potentially contributing to the deeper paradigm shift required for a healthier relationship between humans and the natural world.

Policymakers should carefully consider the value of wildlife, particularly for urban residents. The type of green space matters, and planning policy should extend beyond islands of urban parks to quite extensive, diverse green landscapes ensuring that nature is brought to all residents, turning public spaces into places that prompt sensory contact with nature and provide opportunities to care for nature. Imaginative urban and residential design and wilder cities can provide salient and abundant 'good things of nature' to be noticed and cared for close to where people live.

Creative urban designs could include spaces for green prescribing and provide meaning through using salient nature for wayfinding between nature-rich destinations. For example, the design of hospitals, healthcare premises and care homes could bring nature to people. The routes to these places could be green – for instance, by ensuring that city streets and neighbourhoods contain trees and flowers visible from walkways. Urban 'greenways' that connect parks, transport hubs, schools and shops would improve the environment for all residents, especially if, through good design and information, they encouraged an understanding of the value of noticing nature.

Policy and approaches to managing existing infrastructure can change more swiftly. The results of the research into the power of noticing nature can be incorporated into the design and management of green spaces. Often closely mown, dotted with play areas, they become the visual representation of the term 'amenity greenspace', used in planning to define land that makes a positive contribution to the quality of life. People have become experts at sucking the joy and colour out of nature, somehow making the green beige, sometimes literally through the use of weedkiller around trees where the mower can't reach. Instead, green spaces can be wilder spaces, places for people and wildlife to share. Places that give positive feedback on the state of nature through visible biodiversity. Spaces that become places through creating areas rich in flowers that attract pollinators, such as bumblebees and butterflies. Neatly mown edges of ponds in parks can be left to grow long, creating areas for dragonflies and damselflies to shelter. This creates places for visitors to pause and easily view the wildlife.

Habitat features that promote wildlife such as songbirds, which can be both seen and heard, can be located in prominent places, such as thickets of trees close to paths so that they can be easily explored and experienced by users. To complement habitat management, the impact of more wildlife can be amplified through events, activities, community involvement and technology that creates positive feedback loops to combat the extinction of experience.

Making it real

Excitingly, the principles of nature connection are being applied to strategic planning and the design and management of urban areas. Plymouth aims to be a 'nature city', with the Green Minds project intending to take enterprise and community engagement with nature to the next level, transforming the city's relationship with the natural world.[19] The initiative aims to rewild people, with the pathways to nature connectedness applied in many innovative ways – such as through informing public engagement, communications, the design of community participation activities and green spaces, and several creative commissions. The care and compassion pathways will be activated through community stewardship. The pathways will also inform nature-based land management training and a leadership programme to connect professional audiences with nature. These approaches link through to green social prescribing programmes and design to encourage people to linger and engage with nature-rich areas. The project will help communities and landowners view nature as an asset, increasing the value of nature in policy development via an approach to land management that delivers social and ecological value. The project demonstrates how local government can apply nature connection widely and at scale, placing the relationship between people and nature at the heart of planning and design.

Individuals can also make a difference. In my local city, Jamie Quince-Starkey founded Down to Earth Derby.[20] Everything is connected, and I saw personal connections when I met Jamie – working for Rolls-Royce, a passion for Derby County and a vision for an everyday connection with nature. Jamie was fed up with our relationship with food, the planet and nature. He, too, intuitively saw that our disconnect from nature is the main underlying issue. Just as I'd pursued similar concerns and the need for change through doing research, Jamie pursued them through doing – un-derstanding that changes in day-to-day living are difficult, that the current path and fractured relationship with nature is easier. Where I talked of leverage points, Jamie realised the need to reconnect with nature at every level. Where I talked of reconnection with nature needing to be mainstream, Jamie understood it needed to be realistic. And hopeful. While 25 years in academia has taken me several steps away from some realities, our perspective is shared across a generation.

Down to Earth aims to give people the chance to connect with nature, giving them the tools and confidence to make a difference on their own doorstep. To make a community people want to be part of. Jamie does a great job of sharing his vision and the early days of Down to Earth were successful in engaging others in the community. However, the pandemic enforced a pause, a moment to reach out to someone who could amplify the vision and potentially meet the ambition for change at every level. Eden Project co-founder Sir Tim Smit saw spirit in the project, and helped launch a nature-based vision for the city in the spring of 2022. Business leaders and community figures are engaged by a positive vision of a nature-connected city. Time will tell if that vision becomes a reality, but even now, the story shows how someone with a hopeful and positive vision can make a difference.

Derby is an interesting location for such a project. Derby and the Derwent Valley Mills UNESCO World Heritage Site can be seen as a birthplace of the Industrial Revolution, with the Silk Mill being widely regarded as the site of the world's first modern factory.[21] A factory system that was influential beyond Great Britain. Derbyshire's industry was also built on the exploita-tion of natural resources such as the local coalfields.[22] Derby was also home to the Derby Philosophers, a notable part of the wider Enlightenment movement; the group was founded by Erasmus Darwin and attended by industrialists such as Josiah Wedgwood and other influential figures of the time. Darwin was fascinated by nature and one of the first to forecast the revolutionary conse-quences of the new industry.[23] As we've noted, the consequences of the Industrial Revolution and Enlightenment are the exploitation of natural resources, modern urban living, a broken relationship with nature and a huge increase in wealth. Having taken for free, we are now in debt to nature.

A vibrant natural world has been lost. I'm fortunate to write these words from our converted shed overlooking a field, one set aside and planted for birds. A feeder hangs from the hawthorn a few feet from the window. The feedback on the state of nature is immediate and positive. It is real. Distant flocks of finches stir the January mists as close by blue tits, robins and sparrows visit. Come late April, I'll be hoping to hear the return of the cuckoo, its call a pulse of the landscape, a heart hidden in distant oaks. Travelling the void left by lost brethren, its joyful repeat a reminder of decline. My good fortune in having nature on my doorstep every day is also a reminder that living in a vibrant natural world is alien and a lost right for many. A lost right most don't know they should have, which is why they settle for less.

CHAPTER 15

Designing a Connected Future

Habitats for connection

The policy and systems approach takes a high-level view and can feel a little dry, but informed by the latest research evidence and a focus on a new relationship with nature, presents many opportunities for creative design. Indeed, applying the research evidence to ways to engage people with nature over the past few years has been exciting. There are many creative conversations to be had and voices to hear. For example, I was talking about the pathways to nature connectedness at a National Trust session in 2018, and I was struck by an analogy made by Tony Berry, Visitor Experience Director at the time. He spoke of creating the right habitats to allow a diversity of nature experience to flourish. It made me think about designing 'habitats for connection'.

My original training was in ergonomics, called human factors by some. It is about applying knowledge of people to the design of things for human use, be it a submarine, a kettle or an operating theatre. This human-centred design requires human-centred thinking, systems thinking or design thinking. Whatever the terminology, it has one thing in common – thinking! A sustainable future and closer relationship with nature require new thinking, new ideas and, as we've seen, new systems and even new consumer products and experiences. They must be fit for user needs as before, but balanced with nature's needs and ideally uniting the two. The broken relationship with nature is a result of rigidly human-centred thinking.

With knowledge of people and expertise in applying that knowledge, the human-centred approach of ergonomics can be applied to design in a different way.[1] This is an approach similar to 'nudge theory' popularised by Richard Thaler and Cass Sunstein in 2008. Nudge theory is an enthusiastically discussed and widely adopted concept, that may not be as effective as is often thought.[2] It focuses on our everyday decisions, the poor choices we sometimes make and how our environment can have a bigger impact on behaviour than our character, 'by knowing how people think, we can design choice environments that make it easier for them to choose what is best for themselves, their families, and their society'.[3] Or best for nature.

Thaler is a behavioural economist who saw how people don't always act in their own best interests. We can also add nature's best interest. Like behavioural economics, ergonomics involves understanding human behaviour. Both acknowledge that people are imperfect decision-makers and that the environment can be altered to address this. So, although ergonomics is traditionally about fitting the task to human characteristics to improve safety, performance and wellbeing in the workplace, the associated knowledge and skills could be used to develop an effective approach to changing behaviour through design. This can be informed by the power pathways, noticing and engaging with nature – editing the options available to improve our choices in our everyday decisions that impact human and nature's wellbeing.

An obvious way to design in options and choices is to use the power of affordances. 'Affordance' is a term used by the psychologist James J. Gibson. It originally referred to what the environment 'offers the animal' or the 'complementarity of the animal and the environment'.[4] It is now commonly used to refer to the possibilities for action suggested by the environment; a handle affords holding, a log can afford sitting, a hide affords serious birdwatching, a stick affords sword fighting, pooh sticks and much more! Affordances are direct 'perception–action' processes; just the sight of something can suggest an action, and, as all of the information is available, instructions aren't needed. Just as emotions can arrive from perception, meaningful behaviours can also come directly from perception without too much thought. Clearly, a direct perception–action process is 'fast', such that the environment can provide immediate emotions and meaning. Pathways to nature connectedness can be fast and slow. Nature sometimes presents immediate moments of connection but also suggests, or affords, places to pause and reflect. This relates to the two important dimensions of emotion we balance, joy and calm.

Affordances are powerful, then, providing ways through which the natural landscape talks to us. They can be used to design places and spaces to engage people with nature.[5] Taina Laaksoharju and Erja Rappe from Finland remind us that for children, trees offer multiple and intriguing opportunities for play, such as climbing and den building, that meet different social and personal needs. The use of trees and what they afford becomes more varied over time. As a fundamental step, for affordances and meaningful relationships with nature to emerge, children need spaces where trees are available to connect with – or where the trees can 'speak'. Playgrounds can be designed so that trees are an attraction, with diverse vegetation offering more diverse affordances and experiences for the children. Another example of affordances having a major effect on behaviours is how providing the opportunity for cycling led to the rapid and long-term adoption of the activity as a pro-environmental habit.[6] The design of places can be used as a strong leverage point to form more sustainable habits.[7]

It has also been suggested that affordance theory can fill a blind spot in sense of place research, which tends to focus on social construction of place attachment rather than sensory aspects or meaning arising from the interaction of people and the environment.[8] A sense of place can be generated from being able to 'read' the landscape, or as I'd prefer, hear it speak. Therefore, with an emphasis on meaning and attachment, sense of place is typically seen as 'slow' to evolve and enduring. I think nature connectedness can be seen as akin to a portable sense of place, found within every woodland, passing bird or wherever the good things in nature gather. So, a sense of place in a natural space can be formed from the features of the place, experiences within the place and whatever the individual brings along. Spaces designed for nature connectedness are likely to provide a sense of place, and neither just happen in the head – they're a product of a relationship between the personal and the environment.

Research into affordances in nature also suggests that some people, and children, might need prompting to realise the potential that the environment offers; for many, the natural landscape is mute. Moreover, physical capabilities or cultural factors can lead to people not perceiving opportunities in nature. Designing places that can prompt engagement with nature needs creative thinking and exploration, but the pathways framework can stimulate that thinking. A close relationship with nature starts by affording a place to pause and perceive. A place close to features that stimulate the senses and evoke joy and calm. Places that foster immediate and sustained relationships that bring meaning through function and experiences.

As Gibson noted, affordances can be for 'good or ill', so they should be considered wherever people and nature may interact. There's a relatively new nature reserve close to where I live. Land was purchased, a wildflower meadow set and a parking area provided. Naturally, the hard standing affords parking, the new access affords walking, and a couple of simple benches afford sitting and pausing. As intended, the meadow also affords nesting, but also dog exercising, especially as the parking affords parking by dog owners and professional dog walkers. I've seen people sitting in the cars watching their dogs disappear from sight as they bounce, in what seems a joyful manner, through the long grasses and wildflowers. Professional dog owners release multiple dogs into the meadow. The signs asking for dogs to be kept on leads afford tearing down. The more robust plastic signs afford smashing. The design affords meaning and joy to people and their pets, and it is such a powerful effect that it will take more than a nudge to retrieve the situation. It is a sad result, yet also shows the power of new access to land and affordances to bring meaning. If affordances had been considered from the outset alongside the original aim, which one assumes is to provide new habitats and a place for people to visit and enjoy nature, the end result could be more favourable for human and nature's wellbeing. Affordances happen whether they are intentional or not.

Biophilic design

It is clear that design matters and that our built spaces are also potential habitats for connection. Indeed, given that people spend over 90% of their time indoors, our buildings are important places to foster a closer connection with nature. Biophilic design is a reasonably well-known concept intended to increase connections between people and nature. In 2015, Stephen Kellert and Elizabeth Calabrese published *The Practice of Biophilic Design*, which includes basic principles essential for the successful application of this concept.[9] Some of these principles are highly related to nature connectedness: the need for repeated and sustained engagement with nature, emotional attachment and positive interactions between people and nature. After the principles, the application and practice of biophilic design includes three categories: direct experience of natural features such as natural light and plants; indirect experience of nature through features such as images of nature and natural materials; and spatial features such as open views and places for retreat.

However, these categories are sometimes seen as a mere checklist of materials and physical elements for inclusion. This approach is reflected in research into the benefits of biophilic design, which focuses on the three categories of application: direct, indirect, space and place.[10] When this becomes the focus, the need to foster engagement, emotional attachment and positive interactions with nature can be lost – and we've seen earlier how important active interaction with nature is. The biophilic design principles are an essential element that can be met using the pathways to nature connectedness. Thus, the five pathways and three categories of biophilic design practice can be combined.[11] This approach also allows the evidence base from nature connectedness to help inform and promote biophilic design. For example, earlier in this book, we've seen how direct, rather than indirect, engagement with nature is most important to bring a closer relationship that unites human and nature's wellbeing.

The pathways to nature connectedness provide a framework for direct engagement while embedding the biophilic design principles. It is straightforward to combine and prompt interactions of five different pathway types across the three categories of application. For example, a direct experience of water provides an excellent opportunity for emotional calm and a meaningful place of refuge. Furthermore, pathways and design categories will interact and combine; a place to care for nature can facilitate direct and sensory experience through creating more nature.

As we've seen earlier from the research evidence and affordances, design to foster a closer relationship with nature cannot be passive. The spaces and features created must be engaged with. Moving from the design of a space with certain elements to the behaviour of people in that space requires a different approach, especially when there's a need to evoke emotions and bring meaning. Hence, although good design can influence behaviours, there may well be a need for guidance on how users might enjoy and use spaces, be they biophilic buildings or play areas. Such guidance can also overcome assumptions, such as the idea that nature connection comes from knowledge, learning or passive time in nature. A biophilic workplace may benefit from guidance on break taking and wellbeing programmes that engage people with the nature around them. A biophilic school may benefit from guidance on extracurricular activities, or better still, a curriculum that activates the pathways to nature connectedness. Otherwise, a biophilic space could become what nature is for many, just a background for work or learning – because most people do not notice nature. So, looking into the future of building design, nature must be salient, and people may need to be prompted or even provoked to engage. Design needs to demand attention and use the power of affordances to encourage interaction. Now, in thinking about attention, we draw towards the end of this book, close to where we began.

Our technological future

Earlier in the book, we considered the battle for attention – how most people don't notice nature, and our ability to pay attention is decreasing. Yet, as we've seen, noticing nature is key to a close relationship with it. Our ability to pay attention is declining due to the way we lead our technology-rich lives, especially in those who are more distant from nature.[12] It's not just attention. In the same way that nature affects our emotions, so does technology. It transmits, amplifies and shares emotions at pace, changing behaviour. The rewards of a swipe of a finger across a screen can replace the reward of the next step through a vibrant landscape. As the natural world becomes plainer, people will seek richer lives by migrating to the virtual realm with its 'breath-taking nature'.[13]

Technology shapes and defines us. We are formed by technology and cannot survive without it.[14] Some argue that our broken relationship with nature does not stem from the globalisation of Western values, industrialisation or capitalism; it is 'a consequence of the evolutionary success of an exceptionally rapacious primate' and its growing population's consumption of new technology.[15] And there is more enticing technology to come. Yet, for all our technological advances, the human animal has so far remained the same. Charles Darwin showed that humans are animals; we do not progress and we have the same fate as other animals. Humans serve evolutionary success, not science or truth, but humanists believe we can control the environment and flourish.[16] But to believe the future of the planet is in human hands is to believe in the supremacy of humans over nature, the very source of the broken relationship. Perhaps this paradox is unsolvable.

However, as technology defines our lives, technology could be used for a reconnection with nature. Augmented reality (AR) can bring sterile landscapes back to life, showing us the joy of what was and what can be. Earlier in the book I used the writing of Richard Jefferies to envision the vibrant landscape of the nineteenth century. Wouldn't it be wonderful to witness such scenes? AR provides the opportunity to take a journey from the present, back 100 then 1,000 years to see how our increasingly sterile landscapes were once alive. This would provide an antidote to shifting baseline syndrome, where each generation thinks the current state of nature is the norm. This could be done with AR headsets, but an AR window that overlays the past upon the present would be most powerful. A window on a popular route, where many walk past to enjoy the landscape of today, but where they can witness a truly natural landscape. Of course, even if this were technically possible, a rewilded landscape would be a better solution!

Technology could also magnify each of the pathways to nature connectedness – provide ways to tune into the sounds and sights of nature, to sense the unseen. As tracks in snow record and condense hours of activity, technology can compress hours of activity into minutes, allowing us to see the lives that wildlife leads. Evoking, sharing and amplifying the emotions felt. Creating meaning for the signs and cycles of nature, celebrating wildlife, setting the scene for big stories full of wild characters. Highlighting the interconnectedness of life so that the majority feel part of nature.

When we considered our broken relationship with nature at the start of this book, there were many examples given; most of our actions impact the natural world in some way or other, directly or indirectly. One at times very direct way is via our pets. Cats and dogs consume resources and disturb and kill wildlife, yet they are valued parts of many families. Technology could replace pets. If this seems far-fetched, some propose that in the future we, the technological apes, will have virtual children.[17] Social and companion robots are already here, designed to provide people with emotional and social interactions.[18] They are often animals such as cats and dogs, or even seals, and have been shown to bring positive emotions and reduced depression in dementia.[19] Although there has been a focus on their use in caring for older adults, there are wider opportunities for social and companion robots as technology develops.

Virtual assistants, such as Alexa, are familiar in homes; they could have legs, fur and appealing faces. Just as dogs have been selectively bred to have more human-like expressions, consumer products are designed to appeal to our emotions and even become meaningful parts of our lives.[20] It is easy to imagine a digital assistant being with us and helping in the garden, following the family further afield, but happy home alone. Accompanying us on walks, wayfinding and introducing points of interest based on their understanding of our likes and dislikes. Our mobile assistant might offer, 'You read a book by Richard Jefferies – well, did you know we're passing where he was born? It'll only take a few minutes to go and take a look.' That assistant could take us out into nature at the best times to see it and enjoy it – with knowledge of sightings, how to carefully approach, and prompting us to pause and enjoy the millions of birds no longer falling victim to our feline friends. The opportunities are endless. A future where pets and meat are artificial, but a love of nature is real. Technology that takes us out into a vibrant natural world, dragging us away from ever more immersive virtual worlds. Engaged in the battle for attention, bringing joy and meaning to people, fighting to save nature. Perhaps, where humans have failed, robots and AI will succeed?

This technological vision may appear ridiculous, but as described by Yuval Noah Harari in *Homo Deus: A Brief History of Tomorrow*, technology will know how we feel and know everything we enjoy and find meaningful.[21] Harvesting from your email, browsing biometric devices and conversations, technology will know each individual better than they know themselves. They will become more trusted than our feelings in delivering the human experiences central to our lives, thus becoming the source of authority and meaning. This may sound fanciful, but we already trust technology with many tasks, and evidence suggests we find AI-synthesised faces more trustworthy.[22]

Your robotic pet or digital assistant will know you so well it will become a vital and trusted part of your life. It will seem to have a better idea of what to do, where to go, what to eat than you do yourself. Could technology and consumerism finally deliver the happy and fulfilled life promised for so long? Whether this technology will be trying to make a profit for their corporation or bring about a sustainable future will perhaps depend on the direction set by the developers, those with the biggest budgets. However, with artificial neural networks and machine learning, the technology will perhaps decide. And we'll follow their trusted advice. Will the AI assistant suggest a trip to recovering and abundant wildlife, or will it create a rich virtual experience to separate you from your allocated digital currency? A digital wilderness we might explore as we wander and wave our arms from our airtight carbon-neutral homes. A wilderness we will never map or understand. A technology that mimics nature, eluding the control we thought we had. Or perhaps we will bioengineer in, or out, the love and need for nature.[23]

Ecomodernists believe that the Anthropocene could become a great success if humans use their technological prowess to protect the natural world.[24] For example, some technology companies aim to resurrect versions of long-extinct species, such as the mammoth, to rewild the Arctic.[25] Others suggest this is madness, a humanist promise of humanity's ability to control that prevents action to save the natural world.[26] The $15 million investment in resurrecting the mammoth could instead be targeted at innovative projects to improve the human–nature relationship. One view set out by James Lovelock is that the Anthropocene is an evolutionary product of life on Earth.[27] And from that, 'electronic life' will evolve through a process already in progress. Lovelock suggests that this electronic life may well aim to keep the planet habitable, as climate warming would be destructive for it too. Organic life will help keep the planet cool. This transition to prioritising sustainability over continual expansion is proposed as a reason why aliens have not visited the Earth despite the age and vastness of the universe. There comes a point where technological solutions cannot treat the symptoms of perpetual growth.[28]

Advanced and all-knowing technology may seem to be the route to the humanist dream of power, happiness and immortality, but it may bring an end to humanism. Just as humanism brought a human-centric worldview and sidelined nature, 'dataism' as Harari calls it may sideline humans. Humans would no longer be the planet's most intelligent life form; indeed to AI, humans will be as slow as we find plants while they forge ahead, creating a sustainable future for themselves. The higher power is not nature, a god or even wealth, but data and AI.[29] This new technology might decide human happiness is not essential, that we are not important, and we will become just another animal, treated as we treat animals now. Or we may be like pets.

These visions of a potential future may seem outlandish, but so would our modern world to the hunter-gatherers. Trying to imagine such a future provides some perspective for the present. But what matters, and matters urgently, is the here and now. And that technological future is currently being engineered – offering a new paradigm for us to inhabit, an immersive virtual experience of what all humans once had in the natural world. We are already immersed in online life. We will continue to be woven deeper into the technology that defines us. Virtual experiences will provide self-expression and shared meaning, monetising our identities and virtual pets in a trillion-dollar market.[30] Our stolen immersive and fulfilling relationship with nature will be buried and shifted into a virtual world, created in a distant Silicon Valley. Or we can be woven into a vibrant natural world we can slowly help restore, in our gardens, community spaces and parks, inviting nature in. An immersive world we celebrate and share. When presented with a vision for the future, it is perhaps wise to ask whose worldview we are looking at and what intentions lie behind it.

In his thoughts on humans and other animals, the philosopher John Gray tells us that humans are governed by the needs of the moment and will have the same fate as other animals. Evolution and naturalism tell us that human life is no more meaningful than that of ants. We live under the humanistic illusion of being special and in control. He concludes that the human animal cannot do without purpose and asks, 'Can we not think of the aim of life as being simply to see?'[31] We've established that although simply seeing and noticing is key in reconnecting with nature, it is unpopular. This philosophical wormhole and our attempts to work out our destiny is very human. It seems informative, but to me, as a non-philosopher, it doesn't appear to offer a solution. Just an inescapable conundrum: we must be special to survive, but as part of nature, we cannot be special.

However, for most people, the immediate situation is real. And we've seen that the biodiversity crisis is real, that the broken relationship with nature is real, as are the latent connections and science of nature connectedness. Philosophy aside, people like to believe they have a meaningful life, and they seek happiness. So realising that a close relationship with nature can help provide these things would seem to offer a real story for a sustainable future.

Hope

On a January day, unsure of the final section of this chapter and book, I walked. I walked into a pocket of woodland, houses and shops close by. The lake, brown from the run-off of recent rains, was dotted with mallards and ripples from light rain. Walking just 50 yards, I felt 'within' and paused as birds sang, stepped on again, then remembering the power of pausing and tuning in to the birdsong, I stood. As I'd stood in that woodland a decade before:

> my closest wood, a place I've known since
> childhood. A wood that became smaller as I grew,
> but has regrown as I've learnt how to see with the
> fascination of a child … I stood still, woven by
> the birdsong into the calm of the hollows, sensing
> the proximity of the trees about me.

> A robin explored the woods with sound while a
> blackbird lost its darkness in the holly, and each
> step became another line in the story of the day.
> The sunken stream carried a conveyor of light
> back to the heart of the wood to be digested
> and returned black. The hollows air, its earth, its
> trees seemed liquid, and the water drowned the
> song and drops tapped from tree to ground. The
> sound of movement and peace of still wrapped
> my progress such that I moved with the wood.

> Singing birds emitted spheres of sound,
> expanding invisible as ripples on water,
> disturbed by the trees. I imagined how beautiful
> they might be if time slowed and they grew,
> wrapping around everything to fill all space
> where there is air. I saw the blackbird cutting
> across the sphere of its mate's song, stirring
> the sound with its beat, ripping through it and
> adding the whisper of its wings.

> Returning, my footsteps shone where they had
> leached moisture to the surface of the compressed
> leaf mould, and I left the wood trailing my human
> form as a serpent shedding its skin.

These simple moments in a small woodland close to home became very special. Each tree and each bird provide a sense of wilderness within our nature-depleted landscape. I see trees as the ever-present foundations of a new relationship with nature, birds as our hope – a thread back to the dinosaurs, angels from an extinction. We should listen to their calls. Take flight and break free from centuries of fractured thinking. Yet even now, with research evidence supporting the importance of these moments with nature, I still need to remind myself to pause, to let the woodland suggest a place to stand. To be harboured and tuned into the simple vibrance of birdsong and the light in the leaves. And when I do, I'm always alone. Sometimes there are other visitors walking by. In the woodland without binoculars or by the river without a fishing rod, my standing must appear odd, perhaps suspicious. I've been asked if I've lost my dog! Simple contemplation of nature is not mystical, simply a 'willing surrender to never-returning moments'.[32] We must always be moving or doing. Our current paradigm of living is often busy, technical, consuming, built and grey, with just a few minutes a day spent in green spaces. Yet most people wish modern life could be more straightforward and that the pace of life could slow.[33]

I found my reconnection with nature in simple places close to home. It was a reality that was a catalyst for a decade of research into the human–nature relationship and how to improve it. Yet the approaches to reconnecting with nature are fragmented and at an early stage. There's still a tendency to overlook evidence and fall back on assumptions such as increasing knowledge or simply getting people outdoors. However, it is encouraging that there is increasing recognition that humans and nature are connected, and there are global calls for a new relationship with nature. Yet, despite the theory, we still separate humans from nature in practice, every day across the public realm, thereby reinforcing the current fractured worldview.

The broken relationship between humans and the rest of nature is clear to see through the impact of human actions on the natural habitat and the widespread loss of wildlife. It has happened and continues to happen. Yet millions of people say they love nature and want to protect it. Together through this book, we've travelled back to when humans started to loosen their close bond with nature as they began to farm the land. Finally they saw it snatched away centuries later with the birth of modern detached and individualistic thinking, where the rational mind was seen as the source of truth, and people started to break free from their ties to the natural world. A dualistic rather than relational worldview dominated where human reason, autonomy and science met our needs and created new desires. Humans became addicted to themselves and their technologies.

The emotions at the heart of any relationship were viewed as irrational and were suppressed, before being exploited for commercial gain as our attention was expertly captured. We've seen that growth in measures of modern consumer living such as wealth and smartphone usage are linked to a more distant relationship with nature – those countries with the most growth being the least connected to the natural world. Nature declined, as did our engagement with it in a spiral of disconnection as human needs became dominant. Controlled, dissected, exploited and feared, nature disappeared from our landscapes and our lives. It has even disappeared from our understanding of wellbeing, such that there is a need to prove that people benefit from the habitat of their evolution. We became alien in a new landscape of our own making. An embedded relationship stolen by fractured thinking, a right and necessity many don't realise is missing.

Yet there is hope, in the latent connections between humans and nature, real but overlooked and sometimes denied. Connections within and beyond us, such that our skin is an interface rather than a barrier. How our bodies, if not our rational and conditioned minds, can sense when nature is close by. The vibrancy of nature flows and soothes and manages our moods. We're primed and ready to love nature.

This relationship is encapsulated by the science of nature connectedness, which allows us to put figures on the contributions a close relationship with nature makes to feeling good, functioning well and caring for nature. This reality is captured by 'one health' models that unite human and nature's wellbeing. We've seen that when people are prompted to take a moment and are shown the pathways, they can take steps to a closer relationship with nature. And that those seeking to promote reconnection with nature could learn from the commercial world, applying the power of emotion and meaning more widely and at deeper leverage points within our social structures, from education to urban design and health.

For a new relationship with nature and a sustainable future, there's a need for a culture of connection across the public realm alongside a richer, more vibrant and accessible natural world. A nature-positive society that celebrates the meaning and joy of nature and has a lifelong connection with it. Such a society would need new relational policymaking, creative and shared visions of a close and modern relationship with nature, from the way land and cities are managed to each day at home, in workplaces and at school. A reconnection with nature may be transformative, but it is not transcendental or mystical; it should be an everyday and mainstream reality. A reality that brings meaning to life without the need to chase possessions.

Ultimately, a sustainable future requires a new relational worldview that combines our hunter-gatherer roots with our technological present. A new enlightenment and letting go of the control and use of nature. When I started researching our connection with the rest of nature, I did so because it seemed obvious it was an underlying cause of the environmental crises and a provider of wellbeing. That it takes such persuasion to argue it is a worthwhile path to follow is a further symptom of our broken relationship with nature. Many would sooner focus on technical fixes to the errors in how nature is controlled and used, rather than fixing the underlying failure of the human–nature relationship. Yet a positive story of rekindling lost connections that can bring personal benefits is easier to engage with than a tale of our carbon-heavy ways being taken away. Everyone can find a closer relationship with nature, and when reconnected with nature people will be more prepared to do the difficult things. A positive story of with, rather than without, provides a more compelling vision of a sustainable future. So, let's do what we can to find a friend in nature.

Acknowledgements

Everything is related and this book brings together hundreds of articles by thousands of people – more nodes within the web of the natural world. My thanks to them, and to you, the reader, now part of that network too. Similarly, the success of the research and therefore arrival of the book depends on the vision and creativity of many people at various nature conservation organisations who became early users of the pathways to nature connectedness or supported the research more generally. Likewise, I thank research colleagues in the Nature Connectedness Research Group at the University of Derby for their support, particularly David Sheffield for help getting the research started, Ryan Lumber for the pathways and more recently Carly Butler for bringing new perspectives. Similarly, thank you to Nigel, David and Sarah at Pelagic, for the opportunity and support. And to Sara Magness and Jo Walker for the cover design.

Others have helped more directly with this book, from a simple idea to many hours reading the first draft. My thanks go to Yves Hayaux du Tilly for thoughts around thinking in terms of debt, Eluned Price for information on primate wellbeing, Isabella Tree for reviewing and comments on the section on Knepp, Jake Robinson for reviewing the section on our invisible friends and William Van Gordon for reviewing the section on emptiness. I thank those who reviewed all or parts of the first draft: Leon Richardson, Liz Richardson, Sara Collins, Jo Hardy, Clive Jarrett and Rupert Boddington. A special mention for Robert Little, who sadly and suddenly passed away very soon after kindly completing a full and very useful review. Special thanks to Pauline Little for passing on the manuscript Robert had read and commented on.

Finally, thanks to funky.radio for keeping me on the one while writing about the one.

Notes

Preface

1. *Chuckie Egg* and *Attack of the Mutant Camels* were eight-bit computer games of 1983.

Chapter 1: A Broken Relationship with Nature

1. White, M. P., Elliott, L. R., Grellier, J., et al. (2021). Associations between green/blue spaces and mental health across 18 countries. *Scientific Reports*, 11, 8903. https://doi.org/10.1038/s41598-021-87675-0
2. The People and Nature Survey for England: Monthly indicators with specific weight for April 2020–March 2021 (Official Statistics). (n.d.). GOV.UK. https://www.gov.uk/government/statistics/the-people-and-nature-survey-for-england-monthly-indicators-with-specific-weight-for-april-2020-march-2021-official-statistics
3. Biodiversity indicators | Natural History Museum. (n.d.). https://www.nhm.ac.uk/our-science/data/biodiversity-indicators.html; Briggs, H. (2021, October 9). Biodiversity loss risks "ecological meltdown" - scientists. BBC News. https://www.bbc.co.uk/news/science-environment-58859105
4. Richardson, M. (2021). Did Lockdown Spring Bring a Lasting Connection to Nature? Finding Nature. https://findingnature.org.uk/2021/07/15/did-lockdown-spring-bring-a-lasting-connection-to-nature/
5. Kesebir, S., & Kesebir, P. (2017). A growing disconnection from nature is evident in cultural products. *Perspectives on Psychological Science*, 12(2), 258–69.
6. Twenge, J. M., Campbell, W. K., & Gentile, B. (2013). Changes in pronoun use in American books and the rise of individualism, 1960–2008. *Journal of Cross-Cultural Psychology*, 44(3), 406–15; Twenge, J. M., Campbell, W. K., & Gentile, B. (2012). Increases in individualistic words and phrases in American books, 1960–2008. *PloS One*, 7(7), e40181; Richardson, M. (2022,). Me, myself and nature. Finding Nature. https://findingnature.org.uk/2022/08/31/nature-versus-me/
7. Eckersley, R. (2000). The mixed blessings of material progress: diminishing returns in the pursuit of happiness. *Journal of Happiness Studies*, 1(3), 267–92; Arnett, J. J. (2002). The psychology of globalization. *American Psychologist*, 57(10), 774.

8. António Guterres, Secretary-General of the United Nations, in United Nations Environment Programme (2021). *Making Peace with Nature: A Scientific Blueprint to Tackle the Climate, Biodiversity and Pollution Emergencies.*

9. Legagneux, P., Casajus, N., Cazelles, K., et al. (2018). Our house is burning: discrepancy in climate change vs. biodiversity coverage in the media as compared to scientific literature. *Frontiers in Ecology and Evolution*, 5, 175.

10. WWF (2022). *Living Planet Report*, https://livingplanet.panda.org/en-GB/; Bar-On, Y. M., Phillips, R., & Milo, R. (2018). The biomass distribution on Earth. *Proceedings of the National Academy of Sciences*, 115(25), 6506–11; Carrington, D. (2018, May 22). Humans just 0.01% of all life but have destroyed 83% of wild mammals – study. *The Guardian*. https://www.theguardian.com/environment/2018/may/21/human-race-just-001-of-all-life-but-has-destroyed-over-80-of-wild-mammals-study

11. Brondizio, E. S., Settele, J., Díaz, S., & Ngo, H. T. (2019). *Global Assessment Report on Biodiversity and Ecosystem Services of the Intergovernmental Science-Policy Platform on Biodiversity and Ecosystem Services.*

12. Bar-On, Y. M., Phillips, R., & Milo, R. (2018). The biomass distribution on Earth. *Proceedings of the National Academy of* Sciences, 115(25), 6506–11; Carrington, D. (22 May 2018). Humans just 0.01% of all life but have destroyed 83% of wild mammals – study. The Guardian; The Guardian. https://www.theguardian.com/environment/2018/may/21/human-race-just-001-of-all-life-but-has-destroyed-over-80-of-wild-mammals-study

13. Fanning, A. L., O'Neill, D. W., Hickel, J., & Roux, N. (2021). The social shortfall and ecological overshoot of nations. *Nature Sustainability*, 1–11; https://sustainabilitycommunity.springernature.com/posts/draft

14. Hay festival? Action now for species-rich grasslands: https://www.plantlife.org.uk/application/files/2315/3087/2058/Grasslands_action_plan_-_Plantlife.pdf; The State of the UK's Birds 2020 (2020). https://www.bto.org/sites/default/files/publications/state-of-uk-birds-2020-report.pdf. McKie, R. (10 October 2021). Nearly half of Britain's biodiversity has gone since industrial revolution. *The Guardian*. https://www.theguardian.com/environment/2021/oct/10/nearly-half-of-britains-biodiversity-has-gone-since-industrial-revolution;   Analysis   warns global biodiversity is below "safe limit" ahead of COP 15 (n.d.). www.nhm.ac.uk.   And   https://www.nhm.ac.uk/discover/news/2021/october/analysis-warns-global-biodiversity-is-below-safe-limit.html. Galbraith, P. (2022). *In Search of One Last Song: Britain's Disappearing Birds and the People Trying to Save Them*. London: William Collins.

15. Ceballos, G., Ehrlich, P. R., & Dirzo, R. (2017). Biological annihilation via the ongoing sixth mass extinction signaled by vertebrate population losses and declines. *Proceedings of the National Academy of Sciences*, 114(30), E6089–96.; IPBES (2019). *Summary for Policymakers of the*

Global Assessment Report on Biodiversity and Ecosystem Services of the Intergovernmental Science-Policy Platform on Biodiversity and Ecosystem Services. Bonn, Germany: IPBES Secretariat. https://www.ipbes.net/global-assessment-report-biodiversity-ecosystem-services

16. Jefferies, R. (1883). *Nature Near London*. London: Chatto and Windus.
17. Jefferies, R. (1889). *Field and Hedgerow*. London: Longmans, Green and Company.
18. Landspiracy: is a rewilding revolution inevitable? (n.d.). Theecologist.org. https://theecologist.org/2021/aug/12/landspiracy-rewilding-revolution-inevitable
19. Richardson, M. (2021). A Nation of Nature Lovers? Finding Nature. https://findingnature.org.uk/2021/05/04/a-nation-of-nature-lovers/ Roundup®. (n.d.). Love the Garden. https://www.lovethegarden.com/uk-en/roundup
20. Home And Garden Pesticides Market Report, 2022–2030. (n.d.). www.grandviewresearch.com. https://www.grandviewresearch.com/industry-analysis/home-garden-pesticides-market
21. Lockdown sees surge in demand for artificial grass. (2020, June 29). www.buildersmerchantsnews.co.uk.https://www.buildersmerchantsnews.co.uk/Lockdown-sees-surge-in-demand-for-artificial-grass/49751; Kaminski, I. (2019, August 2). Turf it out: is it time to say goodbye to artificial grass? *The Guardian*. https://www.theguardian.com/cities/2019/aug/02/turf-it-out-is-it-time-to-say-goodbye-to-artificial-grass
22. Francis, R. A. (2018). Artificial lawns: environmental and societal considerations of an ecological simulacrum. *Urban Forestry & Urban Greening*, 30, 152–6.
23. Ro, C. (2020). Why litter is surging as lockdowns ease. www.bbc.com. https://www.bbc.com/worklife/article/20200610-why-are-parks-full-of-litter-as-lockdown-eases; Rowe, M. (2019, December 11). Britain's growing litter problem: why is it so bad and how to take action. Countryfile.com. https://www.countryfile.com/how-to/outdoor-skills/britains-growing-litter-problem-why-is-it-so-bad-and-how-to-take-action/; Brighton beachgoers dump 300 tonnes of rubbish since June. (2021, August 25). BBC News. https://www.bbc.co.uk/news/uk-england-sussex-58326678
24. Kellert, S. R., & Wilson, E. O. (1993). *The Biophilia Hypothesis*. Island Press, Washington, DC.
25. Eckersley, R. (2000). The mixed blessings of material progress; Eisenstein, C. (2013). *The Ascent of Humanity: Civilization and the Human Sense of Self*. Berkeley, CA: North Atlantic Books; Lent, J. R., & Capra, F. (2017). *The Patterning Instinct: A Cultural History of Humanity's Search for Meaning* (p. 409). Amherst, NY: Prometheus Books.
26. Maddison, A. (1995). *Monitoring the World Economy, 1820–1992* (p. 238). Paris: Development Centre of the Organisation for Economic Co-operation and Development.
27. Wiedmann, T., Lenzen, M., Keyßer, L. T., & Steinberger, J. K. (2020). Scientists' warning on affluence. *Nature communications*, 11(1), 1–10.

28. Pilisuk, M. (2001). Ecological psychology, caring, and the boundaries of the person. *Journal of Humanistic Psychology*, 41(2), 25–37.

29. Fanning, A. L., O'Neill, D. W., Hickel, J., & Roux, N. (2021). The social shortfall and ecological overshoot of nations. *Nature Sustainability*, 1–11; Community, S. N. S. (2021, November 17). Charting the social shortfall and ecological overshoot of nations. Springer Nature Sustainability Community. https://sustainabilitycommunity.springernature.com/posts/draft

30. fao. (2018). SOFO 2018 – The State of the World's Forests 2018. www.fao.org. http://www.fao.org/state-of-forests/en/

31. Sharma, A., Kumar, V., Shahzad, B., et al. (2019). Worldwide pesticide usage and its impacts on ecosystem. *SN Applied Sciences*, 1(11), 1–16; Two-thirds of the world's longest rivers no longer run free. (n.d.). www.science.org. https://www.science.org/news/2019/05/two-thirds-world-s-longest-rivers-no-longer-run-free

32. Lopez, B. H. (2001). *Arctic Dreams: Imagination and Desire in a Northern Landscape*. New York: Vintage.

33. Marlowe, F. W. (2005). Hunter-gatherers and human evolution. *Evolutionary Anthropology: Issues, News, and Reviews*, 14(2), 54–67; Schlebusch, C. M., Malmström, H., Günther, T., et al. (2017). Southern African ancient genomes estimate modern human divergence to 350,000 to 260,000 years ago. *Science*, 358(6363), 652–5.

34. Lent, J. R., & Capra, F. (2017). *The Patterning Instinct*.

35. Harari, Y. N. (2016). *Homo Deus: A Brief History of Tomorrow*. New York: Random House.

36. Gray, J. (2002). *Straw Dogs: Thoughts on Humans and Other Animals*. New York: Farrar, Straus and Giroux.

37. Hicks, D. (ed.) (2010). *Ritual and Belief: Readings in the Anthropology of Religion*. Lanham, MD: Rowman Altamira.

38. Naveh, D., & Bird-David, N. (2014). How persons become things: economic and epistemological changes among Nayaka hunter-gatherers. *Journal of the Royal Anthropological Institute*, 20(1), 74–92.

39. An, T. T. M. (2022). Trees are gods: the sanctification of forests in the traditional worldview of the Co Tu people in Vietnam. Вестник Новосибирского государственного университета. Серия: История. Филология, 21(3), 151–61.

40. Bird-David, N. (1999). 'Animism' revisited: personhood, environment, and relational epistemology. *Current Anthropology*, 40(S1), S67–91.

41. Bird-David, N. (1999). 'Animism' revisited.

42. Descola, P. (2013). *Beyond Nature and Culture*. Cambridge: University of Chicago Press; Bird-David, N. (1990). The giving environment: another perspective on the economic system of gatherer-hunters. *Current Anthropology*, 31(2), 189–96.

43. Campbell, J., & Moyers, B. (2011). *The Power of Myth*. New York: Anchor.

44. Galetti, M., Moleón, M., Jordano, P., et al. (2018). Ecological and evolutionary legacy of megafauna extinctions. *Biological Reviews*, 93(2),

845–62. Feeney, J. (2019). Hunter-gatherer land management in the human break from ecological sustainability. *The Anthropocene Review*, 6(3), 223–42. Saltré, F., Rodríguez-Rey, M., Brook, B. W., et al. (2016). Climate change not to blame for late Quaternary megafauna extinctions in Australia. *Nature Communications*, 7(1), 1–7.

45.   Holdaway, R. N., Allentoft, M. E., Jacomb, C., et al. (2014). An extremely low-density human population exterminated New Zealand moa. *Nature Communications*, 5(1), 1–8.

46.   Worm, B., & Paine, R. T. (2016). Humans as a hyperkeystone species. *Trends in Ecology & Evolution*, 31(8), 600–607.

47.   Lovelock, J. (2007). *The Revenge of Gaia: Why the Earth is Fighting Back and How We Can Still Save Humanity* (Vol. 36). London: Penguin.

48.   Graeber, D., & Wengrow, D. (2021). *The Dawn of Everything: A New History of Humanity*. Penguin UK.

49.   Lent, J. R., & Capra, F. (2017). *The Patterning Instinct*; Sahlins, M. (1972). *Stone Age Economics*. New York: Aldine de Gruyter; Sahlins, M. (2006). The original affluent society. In J. Solway (ed.) *The Politics of Egalitarianism: Theory and Practice* (p.79). New York: Bergham Books.

50.   Sahlins, M. (1972). *Stone Age Economics*; Graeber, D., & Wengrow, D. (2021). *The Dawn of Everything*.

51.   MacKerron, G., & Mourato, S. (2013). Happiness is greater in natural environments. *Global Environmental Change*, 23(5), 992–1000; Mears, M., Brindley, P., Barrows, P., et al. (2021). Mapping urban greenspace use from mobile phone GPS data. *PLoS ONE*, 16(7), e0248622.

52.   Huang, T. L. (2021). Restorative experiences and online tourists' willingness to pay a price premium in an augmented reality environment. *Journal of Retailing and Consumer Services*, 58, 102256.

53.   Lent, J. R., & Capra, F. (2017). *The Patterning Instinct* (p. 409); Testart, A., Forbis, R. G., Hayden, B., et al. (1982). The significance of food storage among hunter-gatherers: residence patterns, population densities, and social inequalities [and comments and reply]. *Current Anthropology*, 23(5), 523–37; Dilworth, C. (2010). *Too Smart for Our Own Good: The Ecological Predicament of Humankind*. Cambridge: Cambridge University Press.

54.   Diamond, J. (2013). *Guns, Germs and Steel: A Short History of Everybody for the Last 13,000 Years*. New York: Random House.

55.   Morris, I. (2022). *Geography Is Destiny: Britain and the World, a 10,000 Year History*. London: Profile.

56.   Eisenstein, C. (2013). *The Ascent of Humanity*.

57.   Harari, Y. N. (2016). *Homo Deus*.

58.   Ingold, T. (2000). *The Perception of the Environment: Essays on Livelihood, Dwelling and Skill*. London: Routledge.

59.   Naveh, D., & Bird-David, N. (2014). How persons become things: economic and epistemological changes among Nayaka hunter-gatherers. *Journal of the Royal Anthropological Institute*, 20(1), 74–92.

60.   Piaget, J. (1929). *The Child's Conception of the World*. London: Routledge and Kegan Paul.

61.   Okanda, M., Taniguchi, K., Wang, Y., & Itakura, S. (2021). Preschoolers' and adults' animism tendencies toward a humanoid robot. *Computers in Human Behavior*, 118, 106688; Merewether, J. (2019). Listening with young children: enchanted animism of trees, rocks, clouds (and other things). *Pedagogy, Culture & Society*, 27(2), 233–50.

62.   Barry, J. (2007*). Environment and Social Theory*. London: Routledge; Harari, Y. N. (2016). *Homo Deus*.

63.   Donner, S. D. (2011). Making the climate a part of the human world. *Bulletin of the American Meteorological Society*, 92(10), 1297–302.

64.   Lent, J. R., & Capra, F. (2017). *The Patterning Instinct: A Cultural History of Humanity's Search for Meaning* (p. 409). Amherst, NY: Prometheus Books; Russell, J. R. (1982). *Zoroastrianism in Armenia*. University of London, School of Oriental and African Studies.

65.   Moazami, M. (2005). Evil animals in the Zoroastrian religion. *History of Religions*, 44(4), 300–17; Daryaee, T. (2019). Honey: a demonic food in Zoroastrian Iran? *Studia Litteraria Universitatis Iagellonicae Cracoviensis*, 14, 53–7.

66.   Barr, J. (1985). The question of religious influence: the case of Zoroastrianism, Judaism, and Christianity. *Journal of the American Academy of Religion*, 53(2), 201–35; Kay, J. (1989). Human dominion over nature in the Hebrew Bible. *Annals of the Association of American Geographers*, 79(2), 214–32.

67.   Hamilton, C. (2002). Dualism and sustainability. *Ecological Economics*, 42(1–2), 89–99.

68.   Berman, M. (1981). *The Reenchantment of the World*. Ithaca, NY: Cornell University Press.

Chapter 2: The Great Theft

1.   Gellner, E. (1988). *Plough, Book and Sword: The Structure of Human History*. London: Collins Harvill.

2.   Curd, P. (2001). Why Democritus was not a Skeptic. In A. Preus (ed.) *Essays in Ancient Greek Philosophy* (Vol. 6: *Before Plato*) (pp. 149–69). Albany, NY: SUNY Press.

3.   Harari, Y. N. (2016). *Homo Deus: A Brief History of Tomorrow*. New York: Random House.

4.   Bird-David, N. (1999). 'Animism' revisited: personhood, environment, and relational epistemology. *Current Anthropology*, 40(S1), S67–S91; Morris, B. (1981). Hill gods and ecstatic cults: notes on the religion of a hunting and gathering people. *Man in India Ranchi*, 61(3), 203–36.

5.   Hamilton, C. (2002). Dualism and sustainability. *Ecological Economics*, 42(1–2), 89–99.

6.   Eisenstein, C. (2013). *The Ascent of Humanity: Civilization and the Human Sense of Self*. Berkeley, CA: North Atlantic Books.

7.   Gray, J. (2002). *Straw Dogs: Thoughts on Humans and Other Animals*. New York: Farrar, Straus and Giroux.

8.   Niigaaniin, M., & MacNeill, T. (2022). Indigenous culture and nature relatedness: Results from a collaborative study. *Environmental Development*, 44, 100753.

9.   Graeber, D., & Wengrow, D. (2021). *The Dawn of Everything: A New History of Humanity*. Penguin UK.

10.   Hamilton, C. (2002). Dualism and sustainability. *Ecological Economics*, 42(1–2), 89–99.

11.   Gagliano, M., Vyazovskiy, V. V., Borbély, A. A., et al. (2016). Learning by association in plants. *Scientific Reports*, 6(1), 1–9.

12.   Hamilton, C. (2002). Dualism and sustainability.

13.   Bacon, F. (1863). *The New Organon: Or True Directions Concerning the Interpretation of Nature* (Vol. 1). Library of Alexandria; Castillo, M. (2010). Dualism: from Descartes and Bacon to AJNR. *American Journal of Neuroradiology*, 31(2), 199–200; Eisenstein, C. (2013). *The Ascent of Humanity*; Merrill, T. W. (2008). Masters and possessors of nature review of *Discourse on Method* by René Descartes (trans. Richard Kennington); *On Modern Origins: Essays in Early Modern Philosophy* by Richard Kennington; *Descartes: The Life and Times of a Genius* by A. C. Grayling; Descartes: A Biography by Desmond Clarke; *Descartes? Secret Notebook* by Amir Aczel. *New Atlantis*, 19, 91–107.

14.   Tostevin, B. (2014). *The Promethean Illusion: The Western Belief in Human Mastery of Nature*. Jefferson, NC: McFarland; Mason, J. (2004). An Unnatural Order: The Roots of Our Destruction of Nature. New York: Lantern Books.

15.   Paxman, J. (2021). *Black Gold: The History of How Coal Made Britain*. London: William Collins.

16.   Barry, J. (2007). *Environment and Social Theory*. London: Routledge.

17.   Lovelock, J. (2019). *Novacene: The Coming Age of Hyperintelligence*. Cambridge, MA: MIT Press.

18.   O'Hara, K. (2012). *The Enlightenment: A Beginner's Guide*. New York: Simon and Schuster.

19.   Wiedmann, T., Lenzen, M., Keyßer, L. T., & Steinberger, J. K. (2020). Scientists' warning on affluence. *Nature Communications*, 11(1), 1–10; Weber, M. (1968). *Economy and Society: An Outline of Interpretive Sociology* (Vol. 3). New York: Bedminster Press.

20.   Hartig, T. et al. (2011). Health benefits of nature experience: psychological, social and cultural processes. In K. Nilsson, M. Sangster, C. Gallis, et al. (eds) *Forests, Trees and Human Health* (pp. 127–68). Dordrecht: Springer.

21.   Hazarika, L. X. P. (2012). William Wordsworth and nature: an ecological inquiry. Research Spectrum, 3, 23.

22.   Shamsi, M., & Lashkarian, A. (2014). An ecocritical study on William Wordsworth and Nader Naderpour's poetry. *Asian Journal of Multidisciplinary Studies*, 2(9), 132–41.

23.   Priestley, J. (1771). *An Essay on the First Principles of Government, and on the Nature of Political, Civil and Religious Liberty, Including Remarks on Dr. Brown's Code of Education and on Dr. Balguy's Sermon on Church*

Authority. London: J. Johnson; Eisenstein, C. (2013). *The Ascent of Humanity.*

24. Eckersley, R. (2000). The mixed blessings of material progress: diminishing returns in the pursuit of happiness. *Journal of Happiness Studies*, 1(3), 267–92.

25. John, E. O., & Enang, N. R. (2022). Revisiting the discourse on human–nature relationship in African traditional religion and the responses to the environmental change. *Bangladesh Journal of Bioethics*, 13(1), 39–48.

26. Lent, J. R., & Capra, F. (2017). *The Patterning Instinct: A Cultural History of Humanity's Search for Meaning* (p. 409). Amherst, NY: Prometheus Books.

27. Kidner, D. W. (1994). Why psychology is mute about the environmental crisis. *Environmental Ethics*, 16(4), 359–76.

28. Eisenstein, C. (2013). *The Ascent of Humanity.*

29. Lakoff, G., & Johnson, M. (2008). *Metaphors We Live By.* Chicago: University of Chicago Press.

30. Kidner, D. W. (1994). Why psychology is mute about the environmental crisis.

31. Saunders, C. D. (2003). The emerging field of conservation psychology. *Human Ecology Review*, 10(2), 137–49.

32. Thibodeau, P. H., & Boroditsky, L. (2011). Metaphors we think with: the role of metaphor in reasoning. *PLoS ONE*, 6(2), e16782.

33. Lakoff, G. & Johnson, M. (1999). *Philosophy in the Flesh: The Embodied Mind and its Challenge to Western Thought.* New York: Basic Books.

34. Bang, M., Marin, A., Medin, D., & Washinawatok, K. (2015). Learning by observing, pitching in, and being in relations in the natural world. In *Advances in Child Development and Behavior*, 49, 303–13.

35. Kasser, T., & Ryan, R. M. (1996). Further examining the American dream: differential correlates of intrinsic and extrinsic goals. *Personality and Social Psychology Bulletin*, 22(3), 280–7.

36. Harari, Y. N. (2016). *Homo Deus*; Copson, A. (2015). What is humanism? In A. Copson & A. C. Grayling (eds) *The Wiley Blackwell Handbook of Humanism* (pp. 1–33). Chichester: Wiley Blackwell.

37. Harari, Y. N. (2016). *Homo Deus.*

38. Craig, E. (2013). *Concise Routledge Encyclopedia of Philosophy.* London: Routledge.

39. Pinker, S. (2018). *Enlightenment Now: The Case for Reason, Science, Humanism, and Progress.* London: Penguin.

40. Harari, Y. N. (2016). *Homo Deus.*

41. Copson, A. (2015). What is humanism?

42. Gray, J. (2002). *Straw Dogs: Thoughts on Humans and Other Animals.* New York: Farrar, Straus and Giroux.

43. Twenge, J. M., Campbell, W. K., & Gentile, B. (2013). Changes in pronoun use in American books and the rise of individualism, 1960–2008. *Journal of Cross-Cultural Psychology*, 44(3), 406–15; Twenge, J. M., Campbell, W. K., & Gentile, B. (2012). Increases in individualistic

words and phrases in American books, 1960–2008. *PLoS ONE*, 7(7), e40181.

44. Sophiadi, A. (2014). The song remains the same … or not? A pragmatic approach to the lyrics of rock music. In *Major Trends in Theoretical and Applied Linguistics 2* (pp. 125–42). De Gruyter Open Poland.

45. De Mooij, M., & Hofstede, G. (2011). Cross-cultural consumer behavior: a review of research findings. *Journal of International Consumer Marketing*, 23(3–4), 181–92.

46. Kanner, A. D., & Gomes, M. E. (1995). The all-consuming self. In T. Roszak, M. E. Gomes & A. D. Kanner (eds) *Ecopsychology: Restoring the Earth, Healing the Mind* (pp. 77–91). San Francisco: Sierra Club Books; Twenge, J. M., Miller, J. D., & Campbell, W. K. (2014). The narcissism epidemic: commentary on modernity and narcissistic personality disorder. *Personality Disorders: Theory, Research, and Treatment*, (5)2, 227–9.

47. Logan, A. C., & Prescott, S. L. (2022). Planetary health: we need to talk about narcissism. *Challenges*, 13(1), 19.

48. Leopold, A. (1949). *A Sand County Almanac: With Essays on Conservation from Round River*. New York: Ballantine Books.

49. Van Gordon, W., Shonin, E., Diouri, S., et al. (2018). Ontological addiction theory: attachment to me, mine, and I. *Journal of Behavioral Addictions*, 7(4), 892–6.

50. Shonin, E., Van Gordon, W., & Griffiths, M. D. (2013). Buddhist philosophy for the treatment of problem gambling. *Journal of Behavioral Addictions*, 2(2), 63–71; Shonin, E., Van Gordon, W., Singh, N. N., & Griffiths, M. D. (2015). Mindfulness of emptiness and the emptiness of mindfulness. In E. Shonin, W. Van Gordon, N. N. Singh (eds) *Buddhist Foundations of Mindfulness* (pp. 159–78). Cham: Springer.

51. Newland, G., & Cutler, J. W. (eds) (2014). *The Great Treatise on the Stages of the Path to Enlightenment* (Vol. 2). Boston: Snow Lion Publications; Thurman, R. A. (2004). *Seven Virtues for Living Well*. New York: Riverhead Books.

52. Van Gordon, W., Shonin, E., & Griffiths, M. D. (2017). Buddhist emptiness theory: implications for psychology. *Psychology of Religion and Spirituality*, 9(4), 309.

53. Shonin, E., Van Gordon, W., & Griffiths, M. D. (2016). Ontological addiction: classification, etiology, and treatment. *Mindfulness*, 7(3), 660–71.

54. Van Gordon, W., Shonin, E., & Griffiths, M. D. (2017). Buddhist emptiness theory; Shonin, E., Van Gordon, W., & Griffiths, M. D. (2016). Ontological addiction.

55. Eisenstein, C. (2013). *The Ascent of Humanity*.

56. Weintrobe, S. (2021). *Psychological Roots of the Climate Crisis: Neoliberal Exceptionalism and the Culture of Uncare*. Bloomsbury Publishing, USA; Darwin, C. (1871). *The Descent of Man, and Selection in relation to Sex*. John Murray, London.

57. Thompson, E. (2010). *Mind in Life*. Cambridge, MA: Harvard University Press.

58. Gray, J. (2002). *Straw Dogs: Thoughts on Humans and Other Animals*. New York: Farrar, Straus and Giroux.

59. Angle, S. C. (2009). *Sagehood: The Contemporary Significance of Neo-Confucian Philosophy*. Oxford: Oxford University Press.

60. Quoted in Peaslee, D. C. (1998). *Science as a Cultural Expression*. Commack, NY: Nova Publishers.

61. Liu, J. H. (2017). Neo-Confucian epistemology and Chinese philosophy: practical postulates for actioning psychology as a human science. *Asian Journal of Social Psychology*, 20(2), 137–49; Borthrong, J. (1987). Chu Hsi's ethics: Jen and Ch'eng. *Journal of Chinese Philosophy*, 14(2), 161–78.

62. Alam, K., & Halder, U. K. (2018). A pioneer of environmental movements in India: Bishnoi movement. *Journal of Education and Development*, 8(15), 283–7; Khoshoo, T. N. (1999). The dharma of ecology. *Current Science*, 77(9), 1147–53.

63. The Bahá'í Statement on Nature. (n.d.). www.bahai.org. https://www.bahai.org/documents/bic-opi/bahai-statement-nature

64. Armstrong, A. (ed.) (2019). *Spinoza and Relational Autonomy: Being with Others*. Edinburgh: Edinburgh University Press.

65. Gray, J. (2002). *Straw Dogs*.

66. Tresch, J. (2012). *The Romantic Machine: Utopian Science and Technology after Napoleon*. Chicago: University of Chicago Press.

67. Merleau-Ponty, M. (1962). *The Phenomenology of Perception*, trans. C. Smith. London: Routledge and Kegan Paul. (Original work published 1945.)

68. Clark, A. (1997). *Being There* (p. 222). Cambridge, MA: MIT Press; Gallagher, S. (2005). *How the Body Shapes the Mind*, Oxford: Oxford University Press; Borghi, A. M., & Cimatti, F. (2010). Embodied cognition and beyond: acting and sensing the body. *Neuropsychologia*, 48(3), 763–73; Jacob, P. (2012). Embodied cognition, communication and the language faculty. In Y. Coello & A. Bartolo (eds) *Language and Action in Cognitive Neuroscience* (pp. 21–48). London: Psychology Press.

69. Thompson, E. (2010). *Mind in Life*. Cambridge, MA: Harvard University Press.

70. Bateson, G. (2000). *Steps to an Ecology of Mind: Collected Essays in Anthropology, Psychiatry, Evolution, and Epistemology*. Chicago: University of Chicago Press.

71. Charlton, N. G. (2008). *Understanding Gregory Bateson: Mind, Beauty, and the Sacred Earth*. Albany, NY: SUNY Press.

72. BBC (2019). The state of nature, *Costing the Earth*. www.bbc.co.uk/programmes/m0004sj9

73. Ingold, T. (2000). *The Perception of the Environment: Essays on Livelihood, Dwelling and Skill*. London: Routledge.

74.   United Nations. (2022). The 17 Sustainable Development Goals. United Nations. https://sdgs.un.org/goals

Chapter 3: The Technological Ape

1.   Taylor, T. (2010). *The Artificial Ape: How Technology Changed the Course of Human Evolution*. New York: St Martin's Press.
2.   Eisenstein, C. (2013). *The Ascent of Humanity: Civilization and the Human Sense of Self*. Berkeley, CA: North Atlantic Books.
3.   Harari, Y. N. (2014). *Sapiens: A Brief History of Humankind*. New York: Random House.
4.   Abram, D. (2010). *Becoming Animal: An Earthly Cosmology*. New York: Pantheon Books.
5.   Kesebir, S., & Kesebir, P. (2017). A growing disconnection from nature is evident in cultural products. *Perspectives on Psychological Science*, 12(2), 258–69.
6.   Ownership of smartphones in the UK 2008 & 2019. (n.d.). Statista. https://www.statista.com/statistics/956297/ownership-of-smartphones-uk/
7.   Richardson, M., Hussain, Z., & Griffiths, M. D. (2018). Problematic smartphone use, nature connectedness, and anxiety. *Journal of Behavioral Addictions*, 7(1), 1–8.
8.   Harari, Y. N. (2014). *Sapiens*.
9.   Öhman, A., Flykt, A., & Esteves, F. (2001). Emotion drives attention: detecting the snake in the grass. *Journal of Experimental Psychology: General*, 130(3), 466; Soares, S. C., Lindström, B., Esteves, F., & Öhman, A. (2014). The hidden snake in the grass: superior detection of snakes in challenging attentional conditions. *PLoS ONE*, 9(12), e114724.
10.   Blanchette, I. (2006). Snakes, spiders, guns, and syringes: how specific are evolutionary constraints on the detection of threatening stimuli? *Quarterly Journal of Experimental Psychology*, 59(8), 1484–504.
11.   Konijnendijk, C. C. (2018). The forest of fear. In *The Forest and the City* (pp. 37–50). Cham: Springer.
12.   Richardson, L. (1990). Narrative and sociology. *Journal of Contemporary Ethnography*, 19(1), 116–35.
13.   Pluskowski, A. (2005). The tyranny of the gingerbread house: contextualising the fear of wolves in medieval northern Europe through material culture, ecology and folklore. *Current Swedish Archaeology*, 13(1), 141–60.
14.   Jürgens, U. M., & Hackett, P. M. (2017). The Big Bad Wolf: the formation of a stereotype. *Ecopsychology*, 9(1), 33–43.
15.   Prokop, P., Usak, M., & Erdogan, M. (2011). Good predators in bad stories: cross-cultural comparison of children's attitudes towards wolves. *Journal of Baltic Science Education*, 10(4). 229–42.
16.   Milligan, C., & Bingley, A. (2007). Restorative places or scary spaces? The impact of woodland on the mental well-being of young adults. *Health & Place*, 13(4), 799–811; KS1 English: Little Red Riding Hood. (n.d.).

BBC Class Clips Video. https://www.bbc.co.uk/teach/class-clips-video/english-ks1-little-red-riding-hood-index/zmnrqp3

17.   Schama, S. (1995). *Landscape and Memory*. New York: Harper Perennial.

18.   Konijnendijk, C. C. (2018). The forest of fear. In *The Forest and the City* (pp. 37–50). Cham: Springer.

19.   Johnson, F. R., & Carroll, C. M. (1992). 'Little Red Riding Hood' then and now. *Studies in Popular Culture*, 14(2), 71–84.

20.   Cox, D. T., & Gaston, K. J. (2015). Likeability of garden birds: importance of species knowledge & richness in connecting people to nature. *PloS One*, 10(11), e0141505; Ratcliffe, E., Gatersleben, B., & Sowden, P. T. (2013). Bird sounds and their contributions to perceived attention restoration and stress recovery. *Journal of Environmental Psychology*, 36, 221–8; Jorgensen, A., Hitchmough, J., & Dunnett, N. (2007). Woodland as a setting for housing-appreciation and fear and the contribution to residential satisfaction and place identity in Warrington New Town, UK. *Landscape and Urban Planning*, 79(3–4), 273–87; Jerolmack, C. (2008). How pigeons became rats: the cultural-spatial logic of problem animals. *Social Problems*, 55(1), 72–94.

21.   Jürgens, U. M., & Hackett, P. M. (2021). Wolves, crows, and spiders: an eclectic literature review inspires a model explaining humans' similar reactions to ecologically different wildlife. *Frontiers in Environmental Science*, 9, 3.

22.   Cherry, E. (2018). Birding, citizen science, and wildlife conservation in sociological perspective. *Society & Animals*, 26(2), 130–47; Skandrani, Z., Lepetz, S., & Prévot-Julliard, A. C. (2014). Nuisance species: beyond the ecological perspective. *Ecological Processes*, 3(1), 1–12.

23.   *The Sun* (2020). We're facing a seagull apocalypse as lockdown thwarts pest control – and nest-loving bird mites could even infest homes. https://www.thesun.co.uk/news/11500481/seagull-apocalypse-lockdown-pest-control-mites-rentokil/; Simcox, G. (2021, July 19). Man "beats seagull to death with child's spade on a beach in Cornwall." Mail Online. https://www.dailymail.co.uk/news/article-9801163/Man-beats-seagull-death-childs-spade-beach-Cornwall.html

24.   Irwin, A., Geschke, A., Brooks, T. M., et al. (2022). Quantifying and categorising national extinction-risk footprints. *Scientific Reports*, 12(1), 1–10.

25.   Bötsch, Y., Tablado, Z., Scherl, D., et al. (2018). Effect of recreational trails on forest birds: human presence matters. *Frontiers in Ecology and Evolution*, 6, 175; Remacha, C., Delgado, J. A., Bulaic, M., & Perez-Tris, J. (2016). Human disturbance during early life impairs nestling growth in birds inhabiting a nature recreation area. *PloS One*, 11(11), e0166748; Bötsch, Y., Tablado, Z., & Jenni, L. (2017). Experimental evidence of human recreational disturbance effects on bird-territory establishment. *Proceedings of the Royal Society B: Biological Sciences*, 284(1858), 20170846.

26.   PFMA. (2021). NEW: Pet Population 2021. www.pfma.org.uk. https://www.pfma.org.uk/pet-population-2021

27. Morgan, L., Protopopova, A., Birkler, R. I. D., et al. (2020). Human–dog relationships during the COVID-19 pandemic: booming dog adoption during social isolation. *Humanities and Social Sciences Communications*, 7(1), 1–11; Olivia, J. L., & Johnston, K. L. (2021). Puppy love in the time of Corona: dog ownership protects against loneliness for those living alone during the COVID-19 lockdown. *International Journal of Social Psychiatry*, 67(3), 232–42; BBC (12 March 2021). Households 'buy 3.2 million pets in lockdown'. BBC News. https://www.bbc.co.uk/news/business-56362987

28. Tulloch, J. S., Minford, S., Pimblett, V., et al. (2021). Paediatric emergency department dog bite attendance during the COVID-19 pandemic: an audit at a tertiary children's hospital. *BMJ Paediatrics Open*, 5(1); Burford, R. (2021, July 5). One in 5 dog owners who bought "pandemic puppy" consider rehoming pet. Evening Standard. https://www.standard.co.uk/news/uk/pandemic-puppy-dog-rehoming-b944169.html

29. Young, J. K., Olson, K. A., Reading, R. P., et al. (2011). Is wildlife going to the dogs? Impacts of feral and free-roaming dogs on wildlife populations. *BioScience*, 61(2), 125–32.

30. Hughes, J., & Macdonald, D. W. (2013). A review of the interactions between free-roaming domestic dogs and wildlife. *Biological Conservation*, 157, 341–51; Marra, P. P. (2019). The ecological cost of pets. *Current Biology*, 29(19), R955–6.

31. Miller, S. G., Knight, R. L., & Miller, C. K. (2001). Wildlife responses to pedestrians and dogs. *Wildlife Society Bulletin*, 124–32.

32. Banks, P. B., & Bryant, J. V. (2007). Four-legged friend or foe? Dog walking displaces native birds from natural areas. *Biology Letters*, 3(6), 611–13.

33. Quetteville, H. de, & Butcher, A. (2022, May 13). Dog fights: the rise of rage in the countryside. *The Telegraph*. https://www.telegraph.co.uk/news/2022/05/13/dog-fights-rise-rage-countryside/

34. Lord, A., Waas, J. R., Innes, J., & Whittingham, M. J. (2001). Effects of human approaches to nests of northern New Zealand dotterels. *Biological Conservation*, 98(2), 233–40.

35. Doherty, T. S., Glen, A. S., Nimmo, D. G., et al. (2016). Invasive predators and global biodiversity loss. *Proceedings of the National Academy of Sciences*, 113(40), 11261–5.

36. Loss, S. R., Will, T., & Marra, P. P. (2013). The impact of free-ranging domestic cats on wildlife of the United States. *Nature Communications*, 4(1), 1–8.

37. Pirie, T. J., Thomas, R. L., & Fellowes, M. D. (2022). Pet cats (*Felis catus*) from urban boundaries use different habitats, have larger home ranges and kill more prey than cats from the suburbs. *Landscape and Urban Planning*, 220, 104338; Woods, M., McDonald, R. A., & Harris, S. (2003). Predation of wildlife by domestic cats *Felis catus* in Great Britain. *Mammal Review*, 33(2), 174–88; Marra, P. P. (2019). The ecological cost of pets. *Current Biology*, 29(19), R955–6; Trouwborst, A., McCormack, P. C., &

Martínez Camacho, E. (2020). Domestic cats and their impacts on biodiversity: a blind spot in the application of nature conservation law. *People and Nature*, 2(1), 235–50.

38. McDonald, J. L., & Skillings, E. (2021). Human influences shape the first spatially explicit national estimate of urban unowned cat abundance. *Scientific Reports*, 11(1), 1–12.

39. Alexander, P., Berri, A., Moran, D. et al. (2020). The global environmental paw print of pet food. *Global Environmental Change*, 65, 102153.

40. Okin, G. S. (2017). Environmental impacts of food consumption by dogs and cats. *PLoS ONE*, 12(8), e0181301.

41. The truth about cats' and dogs' environmental impact. (n.d.). UCLA. https://newsroom.ucla.edu/releases/the-truth-about-cats-and-dogs-environmental-impact

42. Herrera-Camacho, J., Baltierra-Trejo, E., Taboada-González, P. A., et al. (2017). Environmental footprint of domestic dogs and cats. Preprints, 2017070004.

43. Rushforth, R., & Moreau, M. (2013). Finding your dog's ecological 'pawprint': a hybrid EIO-LCA of dog food manufacturing. *Course Project Report Series SSEBE-CESEM-2013-CPR*, 5.

44. Alexander, P., Berri, A., Moran, D., et al. (2020). The global environmental paw print of pet food. *Global Environmental Change*, 65, 102153.

45. Dogs TV channel: DogTV set to launch to help with stress and behavioural problems. (n.d.). Sky News. https://news.sky.com/story/dogs-tv-channel-dogtv-set-to-launch-to-help-with-stress-and-behavioural-problems-12459735; Patrick, H. (n.d.). Why more companies are going dog friendly. The Conversation. https://theconversation.com/why-more-companies-are-going-dog-friendly-123405

46. Blouin, D. D. (2013). Are dogs children, companions, or just animals? Understanding variations in people's orientations toward animals. *Anthrozoös*, 26(2), 279–94.

47. Martens, P., Su, B., & Deblomme, S. (2019). The ecological paw print of companion dogs and cats. *BioScience*, 69(6), 467–74.

48. Nisbet, E. K., Zelenski, J. M., & Murphy, S. A. (2009). The nature relatedness scale: linking individuals' connection with nature to environmental concern and behavior. *Environment and Behavior*, 41(5), 715–40.

49. White, M. P., Elliott, L. R., Taylor, T., et al. (2016). Recreational physical activity in natural environments and implications for health: a population based cross-sectional study in England. *Preventive Medicine*, 91, 383–8.

50. Fretwell, K., & Greig, A. (2019). Towards a better understanding of the relationship between individual's self-reported connection to nature, personal well-being and environmental awareness. *Sustainability*, 11(5), 1386.

51. Lumber, R., Richardson, M., & Sheffield, D. (2018). The seven pathways to nature connectedness: a focus group exploration. *European Journal of Ecopsychology*, 6, pp. 47–68.

52. Colléony, A., White, R., & Shwartz, A. (2019). The influence of spending time outside on experience of nature and environmental attitudes. *Landscape and Urban Planning*, 187, 96–104.

53. Government Office for Science (2008). Mental capital and wellbeing. GOV.UK. https://www.gov.uk/government/collections/mental-capital-and-wellbeing

54. Government Office for Science (2008). Mental capital and wellbeing: making the most of ourselves in the 21st century. GOV.UK. https://www.gov.uk/government/publications/mental-capital-and-wellbeing-making-the-most-of-ourselves-in-the-21st-century

55. Five Ways to Well-being. (n.d.). Issuu. https://issuu.com/neweconomicsfoundation/docs/five_ways_to_well-being?viewMode=presentation; NHS. (2019). 5 Steps to Mental Wellbeing. NHS.https://www.nhs.uk/mental-health/self-help/guides-tools-and-activities/five-steps-to-mental-wellbeing/

56. Niedzwiedz, C. L., Green, M. J., Benzeval, M., et al. (2021). Mental health and health behaviours before and during the initial phase of the COVID-19 lockdown: longitudinal analyses of the UK Household Longitudinal Study. *Journal of Epidemiological and Community Health*, 75(3), 224–231.

57. Aked, J. (2011). *Five ways to wellbeing: New applications, new ways of thinking.* New Economics Foundation.

58. Patel, V., Saxena, S., Lund, C., et al. (2018). The Lancet Commission on global mental health and sustainable development. *The Lancet*, 392(10157), 1553–98.

59. Barboza, E. P., Cirach, M., Khomenko, S., et al. (2021). Green space and mortality in European cities: a health impact assessment study. *The Lancet Planetary Health*, 5(10), e718–30.

60. Topic: Advertising in the United Kingdom. (2017). Statista. https://www.statista.com/topics/1747/advertising-in-the-united-kingdom/

61. UK government beats Unilever and Sky as biggest UK advertiser of 2020. (n.d.). *The Drum.* https://www.thedrum.com/news/2021/03/23/uk-government-beats-unilever-and-sky-biggest-uk-advertiser-2020

62. Van de Sand, F., Frison, A. K., Zotz, P., et al. (2020). The battle for attention. In *User Experience Is Brand Experience* (pp. 1–16). Cham: Springer.

63. Richardson, M. & Hamlin, I. (2021). Nature engagement for human and nature's wellbeing during the Corona pandemic. *Journal of Public Mental Health*. https://doi.org/10.1108/JPMH-02-2021-0016

64. Hyvärinen, H., & Beck, R. (2018, January). Emotions trump facts: the role of emotions in on social media: a literature review. In *Proceedings of the 51st Hawaii International Conference on System Sciences*. https://doi.org/10.24251/HICSS.2018.226

65. Wahl-Jorgensen, K. (2018). The emotional architecture of social media. In Z. Papacharissi (ed.) *A Networked Self and Platforms, Stories, Connections* (pp. 77–93). New York: Routledge; Brady, W. J., Crockett, M. J., & Van Bavel, J. J. (2020). The MAD model of moral contagion:

the role of motivation, attention, and design in the spread of moralized content online. *Perspectives on Psychological Science*, 15(4), 978–1010.

66. Poels, K., & Dewitte, S. (2019). The role of emotions in advertising: a call to action. *Journal of Advertising*, 48(1), 81–90.

67. The Century of the Self. (2018). www.hgi.org.uk. https://www.hgi.org.uk/resources/delve-our-extensive-library/interviews/century-self

68. Bernays, E. L. (1928). Manipulating public opinion: the why and the how. *American Journal of Sociology*, 33(6), 958–71.

69. Kanner, A., & Gomes, M. (1995). The all-consuming self. In T. Roszak, M. Gomes & A. Kanner (eds) *Ecopsychology* (pp. 77–91). San Francisco: Sierra Club Books.

70. Eckersley, R. (2000). The mixed blessings of material progress: diminishing returns in the pursuit of happiness. *Journal of Happiness Studies*, 1(3), 267–92.

71. Morgan, B. (n.d.). NOwnership, No Problem: An Updated Look At Why Millennials Value Experiences Over Owning Things. Forbes. https://www.forbes.com/sites/blakemorgan/2019/01/02/nownership-no-problem-an-updated-look-at-why-millennials-value-experiences-over-owning-things/?sh=27bc77ac522f

72. Niemietz, K. P. (2021). Left Turn Ahead: Surveying Attitudes of Young People Towards Capitalism and Socialism. *SSRN Electronic Journal*. https://doi.org/10.2139/ssrn.3893595

73. Van de Sand, F., Frison, A. K., Zotz, P., et al. (2020). The battle for attention. In *User Experience Is Brand Experience* (pp. 1–16). Cham: Springer.

74. Kale, S. (2021). Out of style: Will Gen Z ever give up its dangerous love of fast fashion? *The Guardian*. https://www.theguardian.com/fashion/2021/oct/06/out-of-style-will-gen-z-ever-give-up-its-dangerous-love-of-fast-fashion

Chapter 4: Hidden Connections with Nature

1. Richardson, M., Maspero, M., Golightly, D., et al. (2017). Nature: a new paradigm for well-being and ergonomics. *Ergonomics*, 60(2), 292–305; Russell, R., Guerry, A. D., Balvanera, P., et al. (2013). Humans and nature: how knowing and experiencing nature affect well-being. *Annual Review of Environment and Resources*, 38, 473–502; Maller, C., Townsend, M., Pryor, A., et al. (2006). Healthy nature healthy people: 'contact with nature' as an upstream health promotion intervention for populations. *Health Promotion International*, 21(1), 45–54; Maas, J., Verheij, R. A., Groenewegen, P. P., et al. (2006). Green space, urbanity, and health: how strong is the relation? *Journal of Epidemiology & Community Health*, 60(7), 587–92; Barboza, E. P., Cirach, M., Khomenko, S., et al. (2021). Green space and mortality in European cities: a health impact assessment study. *The Lancet Planetary Health*, 5(10), e718–30.

2. McMahan, E. A., & Estes, D. (2015). The effect of contact with natural environments on positive and negative affect: a meta-analysis. *Journal of Positive Psychology*, 10(6), 507–19; Hartig, T., Evans, G. W., Jamner, L. D.,

et al. (2003). Tracking restoration in natural and urban field settings. *Journal of Environmental Psychology*, 23(2), 109–23.

3.   Roszak, T. (1995). Where psyche meets Gaia. In T. Roszak, M. E. Gomes & A. D. Kanner (eds) *Ecopsychology: Restoring the Earth, Healing the Mind* (pp. 1–17). San Francisco: Sierra Club Books; Roszak, T. (1992). *The Voice of the Earth*. New York: Touchstone.

4.   Cameron, R. W. F., Brindley, P., Mears, M., et al. (2020). Where the wild things are! Do urban green spaces with greater avian biodiversity promote more positive emotions in humans? *Urban Ecosystems*. https://doi.org/10.1007/s11252-020-00929-z

5.   Qiu, L., Lindberg, S., & Nielsen, A. B. (2013). Is biodiversity attractive? – On-site perception of recreational and biodiversity values in urban green space. *Landscape and Urban Planning*, 119, 136–46; Fuller, R. A., Irvine, K. N., Devine-Wright, P., et al. (2007). Psychological benefits of greenspace increase with biodiversity. *Biology Letters*, 3(4), 390–4; Douglas, J. W., & Evans, K. L. (2021). An experimental test of the impact of avian diversity on attentional benefits and enjoyment of people experiencing urban green-space. *People and Nature*. https://doi.org/10.1002/pan3.10279

6.   Bravo, J. A., Julio-Pieper, M., Forsythe, P., et al. (2012). Communication between gastrointestinal bacteria and the nervous system. *Current Opinion in Pharmacology*, 12(6), 667–72; Clapp, M., Aurora, N., Herrera, L., et al. (2017). Gut microbiota's effect on mental health: the gut–brain axis. *Clinics and Practice*, 7(4), 131–6.

7.   Mirzaei, M. K., & Maurice, C. F. (2017). The mammalian gut as a matchmaker. *Cell Host & Microbe*, 22(6), 726–7.

8.   Lucas, G. (2018). Gut thinking: the gut microbiome and mental health beyond the head. *Microbial Ecology in Health and Disease*, 29(2), 1548250.

9.   Rees, T., Bosch, T., & Douglas, A. E. (2018). How the microbiome challenges our concept of self. *PLoS Biology*, 16(2), e2005358.

10. Sender, R., Fuchs, S., & Milo, R. (2016). Revised estimates for the number of human and bacteria cells in the body. *PLoS Biology*, 14(8), e1002533.

11. Robinson, J. M., Mills, J. G., & Breed, M. F. (2018). Walking ecosystems in microbiome-inspired green infrastructure: an ecological perspective on enhancing personal and planetary health. *Challenges*, 9(2), 40; Dheilly, N. M. (2014). Holobiont–holobiont interactions: redefining host–parasite interactions. *PLoS Pathogens*, 10(7), e1004093; Van de Guchte, M., Blottière, H. M., & Doré, J. (2018). Humans as holobionts: implications for prevention and therapy. *Microbiome*, 6(1), 1–6.

12. Flandroy, L., Poutahidis, T., Berg, G., et al. (2018). The impact of human activities and lifestyles on the interlinked microbiota and health of humans and of ecosystems. *Science of the Total Environment*, 627, 1018–38.

13. Tonello, L., Gashi, B., Scuotto, A., et al. (2018). The gastrointestinal-brain axis in humans as an evolutionary advance of the root–leaf axis

in plants: a hypothesis linking quantum effects of light on serotonin and auxin. *Journal of Integrative Neuroscience*, 17(2), 227–37.

14.   Robinson, J. M., Breed, M., & Cameron, R. (n.d.). How the trees in your local park help protect you from disease. *The Conversation*. https://theconversation.com/how-the-trees-in-your-local-park-help-protect-you-from-disease-160312

15.   Robinson, J. M., Breed, M., & Cameron, R. (n.d.). How the trees in your local park help protect you from disease; Robinson, J. M., Cando-Dumancela, C., Antwis, R. E., et al. (2021). Exposure to airborne bacteria depends upon vertical stratification and vegetation complexity. *Scientific Reports*, 11(1), 1–16.

16.   Robinson, J. M. (n.d.). Why spending more time in nature could reduce "germaphobia." *The Conversation*. https://theconversation.com/why-spending-more-time-in-nature-could-reduce-germaphobia-163741

17.   Robinson, J. M. (n.d.). Biodiversity loss could be making us sick – here's why. *The Conversation*. https://theconversation.com/biodiversity-loss-could-be-making-us-sick-heres-why-143627

18.   Rees, T., Bosch, T., & Douglas, A. E. (2018). How the microbiome challenges our concept of self. *PLoS Biology*, 16(2), e2005358.

19.   Baxter, M. G., & Croxson, P. L. (2012). Facing the role of the amygdala in emotional information processing. *Proceedings of the National Academy of Sciences*, 109(52), 21180–1.

20.   Kühn, S., Düzel, S., Eibich, P., et al. (2017). In search of features that constitute an 'enriched environment' in humans: associations between geographical properties and brain structure. *Scientific Reports*, 7, 11920.

21.   Kim, M. J., Avinun, R., Knodt, A. R., et al. (2017). Neurogenetic plasticity and sex influence the link between corticolimbic structural connectivity and trait anxiety. *Scientific Reports*, 7, 10959.

22.   Ikei, H., Song, C., & Miyazaki, Y. (2017). Physiological effects of touching wood. *International Journal of Environmental Research and Public Health*, 14(7), 801.

23.   Ikei, H., Song, C., & Miyazaki, Y. (2017). Physiological effects of touching coated wood, 773.

24.   Lee, J., Park, B. J., Tsunetsugu, Y., et al. (2011). Effect of forest bathing on physiological and psychological responses in young Japanese male subjects. *Public Health*, 125(2), 93–100; Park, B. J., Tsunetsugu, Y., Kasetani, T., et al. (2009). Physiological effects of forest recreation in a young conifer forest in Hinokage Town, Japan. *Silva Fennica*, 43(2), 291–301.

25.   Song, C., Igarashi, M., Ikei, H., & Miyazaki, Y. (2017). Physiological effects of viewing fresh red roses. *Complementary Therapies in Medicine*, 35, 78–84.

26.   Mochizuki-Kawai, H., Matsuda, I., & Mochizuki, S. (2020). Viewing a flower image provides automatic recovery effects after psychological stress. *Journal of Environmental Psychology*, 70, 101445.

27. Sudimac, S., Sale, V. & Kühn, S. (2022). How nature nurtures: Amygdala activity decreases as the result of a one-hour walk in nature. *Molecular Psychiatry*, 10.1038/s41380-022-01720-6

28. Koivisto, M., Jalava, E., Kuusisto, L., et al. (2022). Top-down processing and nature connectedness predict psychological and physiological effects of nature. *Environment and Behavior*. https://doi.org/10.1177/00139165221107535

29. Gross, J. J. (2013). Emotion regulation: taking stock and moving forward. *Emotion*, 13, 359.

30. Korpela, K. M., Pasanen, T., Repo, V., et al. (2018). Environmental strategies of affect regulation and their associations with subjective well-being. *Frontiers in Psychology*, 9, 562.

31. Gilbert, P. (ed.) (2005). *Compassion: Conceptualisations, Research and Use in Psychotherapy*. Hove: Routledge; Gilbert, P. (2014). The origins and nature of compassion focused therapy. *British Journal of Clinical Psychology*, 53, 6–41.

32. Richardson, M., McEwan, K., Maratos, F., & Sheffield, D. (2016). Joy and calm: how an evolutionary functional model of affect regulation informs positive emotions in nature. *Evolutionary Psychological Science*, 2(4), 308–20; Richardson, M. (2019). Beyond restoration: considering emotion regulation in natural well-being. *Ecopsychology*, 11(2), 123–9.

33. Kuo, M. (2015). How might contact with nature promote human health? Promising mechanisms and a possible central pathway. *Frontiers in Psychology*, 6, 1093; Marsland, A. L., Pressman, S., & Cohen, S. (2007). Positive affect and immune function. In R. Ader (ed.) *Psychoneuroimmunology* (4th edn). San Diego, CA: Elsevier.

34. Robinson, J. M., Mills, J. G., & Breed, M. F. (2018). Walking ecosystems in microbiome-inspired green infrastructure: an ecological perspective on enhancing personal and planetary health. *Challenges*, 9(2), 40; Dobson, A., Lodge, D., Alder, J., et al. (2006). Habitat loss, trophic collapse, and the decline of ecosystem services. *Ecology*, 87(8), 1915–24.

35. Cameron, R. W. F., Brindley, P., Mears, M., et al. (2020). Where the wild things are! Do urban green spaces with greater avian biodiversity promote more positive emotions in humans? *Urban Ecosystems*. https://doi.org/10.1007/s11252-020-00929-z

36. Dallimer, M., Irvine, K. N., Skinner, A. M., et al. (2012). Biodiversity and the feel-good factor: understanding associations between self-reported human well-being and species richness. *BioScience*, 62(1), 47–55; Fuller, R. A., Irvine, K. N., Devine-Wright, P., et al. (2007). Psychological benefits of greenspace increase with biodiversity. *Biology Letters*, 3(4), 390–4; Wood, E., Harsant, A., Dallimer, M., et al. (2018). Not all green space is created equal: biodiversity predicts psychological restorative benefits from urban green space. *Frontiers in Psychology*, 9, 2320.

37. Methorst, J., Rehdanz, K., Mueller, T., et al. (2021). The importance of species diversity for human well-being in Europe. *Ecological Economics*, 181, 106917.

38. Hepburn, L., Smith, A. C., Zelenski, J., & Fahrig, L. (2021). Bird diversity unconsciously increases people's satisfaction with where they live. *Land*, 10(2), 153.

39. Kaplan, S. (1975). An informal model for the prediction of preference. In E. H. Zube, R. O. Brush and J. G. Fabos (eds) *Landscape Assessment: Values, Perception and Resources* (pp. 92–101). Stroudsburg, PA: Dowden, Hutchinson and Ross; Ulrich, R. S. (1983). Aesthetic and affective response to natural environment. In I. Altman & J. F. Wohlwill (eds) *Behavior and the Natural Environment* (pp. 85–125). Boston, MA: Springer. https://doi.org/10.1007/978-1-4613-3539-9_4

40. Schiebel, T., Gallinat, J., & Kühn, S. (2022). Testing the biophilia theory: automatic approach tendencies towards nature. *Journal of Environmental Psychology*, 79, 101725; Joye, Y., & van den Berg, A. (2011). Is love for green in our genes? A critical analysis of evolutionary assumptions in restorative environments research. *Urban Forestry & Urban Greening*, 10(4), 261–8.

41. Yoshimoto, S., Imai, H., Kashino, M., & Takeuchi, T. (2014). Pupil response and the subliminal mere exposure effect. *PloS One*, 9(2), e90670.

42. Franěk, M., Petružálek, J., & Šefara, D. (2019). Eye movements in viewing urban images and natural images in diverse vegetation periods. *Urban Forestry & Urban Greening*, 46, 126477.

43. Miyazaki, Y., Park, B. J., & Lee, J. (2011). Nature therapy. In M. Osaki, A. K. Braimoh & K. Nakagami (eds) *Designing Our future: Local Perspectives on Bioproduction, Ecosystems and Humanity* (pp. 407–12). Tokyo: United Nations University Press.

Chapter 5: Nature Connectedness

1. Hatty, M. A., Goodwin, D., Smith, L. D. G., & Mavondo, F. (2022, May 12). Speaking of nature: relationships between how people think about, connect with, and act to protect nature. https://doi.org/10.31219/osf.io/wtgy3

2. Schultz, P. W. (2002). Inclusion with nature: the psychology of human–nature relations. In P. Schmuck & W. P. Schultz (eds) *Psychology of Sustainable Development* (pp. 61–78). Boston, MA: Springer.

3. Schultz, P. W. (2001). The structure of environmental concern: concern for self, other people, and the biosphere. *Journal of Environmental Psychology*, 21(4), 327–39.

4. Tam, K. P. (2013). Concepts and measures related to connection to nature: similarities and differences. *Journal of Environmental Psychology*, 34, 64–78.

5. Mayer, F. S., & Frantz, C. M. (2004). The connectedness to nature scale: a measure of individuals' feeling in community with nature. *Journal of Environmental Psychology*, 24(4), 503–15; Nisbet, E. K., Zelenski, J. M., & Murphy, S. A. (2009). The nature relatedness scale: linking individuals' connection with nature to environmental concern

and behavior. *Environment and Behavior*, 41(5), 715–40; Nisbet, E. K., & Zelenski, J. M. (2013). The NR-6: a new brief measure of nature relatedness. *Frontiers in Psychology*, 4, 813.

6. Perrin, J. L., & Benassi, V. A. (2009). The connectedness to nature scale: a measure of emotional connection to nature? *Journal of Environmental Psychology*, 29(4), 434–40.

7. Frantz, C., Mayer, F. S., Norton, C., & Rock, M. (2005). There is no 'I' in nature: the influence of self-awareness on connectedness to nature. *Journal of Environmental Psychology*, 25(4), 427–36.

8. Chang, C.-c., Cox, D. T. C., Fan, Q., et al. (2022). People's desire to be in nature and how they experience it are partially heritable. *PLoS Biology* 20(2), e3001500. https://doi.org/10.1371/journal.pbio.3001500

9. Martin, L., White, M. P., Hunt, A., et al. (2020). Nature contact, nature connectedness and associations with health, wellbeing and pro-environmental behaviours. *Journal of Environmental Psychology*, 68, 101389.

10. Hurly, J., & Walker, G. J. (2019). Nature in our lives: examining the human need for nature relatedness as a basic psychological need. *Journal of Leisure Research*, 50(4), 290–310; Baxter, D. E., & Pelletier, L. G. (2019). Is nature relatedness a basic human psychological need? A critical examination of the extant literature. *Canadian Psychology/ Psychologie canadienne*, 60(1), 21.

11. Lambert, L., Lomas, T., van de Weijer, M. P., et al. (2020). Towards a greater global understanding of wellbeing: a proposal for a more inclusive measure. *International Journal of Wellbeing*, 10(2); Gallup. (2014). How Does the Gallup World Poll Work? Gallup.com https:// www.gallup.com/178667/gallup-world-poll-work.aspx

12. SEI & CEEW (2022). Stockholm+50: Unlocking a Better Future. Stockholm Environment Institute.

13. Ives, C. D., Abson, D. J., von Wehrden, H., et al. (2018). Reconnecting with nature for sustainability. *Sustainability Science*, 13(5), 1389–97.

14. Richardson, M., Hunt, A., Hinds, J., et al. (2019). A measure of nature connectedness for children and adults: validation, performance, and insights. *Sustainability*, 11(12), 3250.

15. Hughes, J., Rogerson, M., Barton, J., & Bragg, R. (2019). Age and connection to nature: when is engagement critical?. *Frontiers in Ecology and the Environment*, 17(5), 265–9; Krettenauer, T., Wang, W., Jia, F., & Yao, Y. (2020). Connectedness with nature and the decline of pro-environmental behavior in adolescence: a comparison of Canada and China. *Journal of Environmental Psychology*, 71, 101348.

16. Piccininni, C., Michaelson, V., Janssen, I., & Pickett, W. (2018). Outdoor play and nature connectedness as potential correlates of internalized mental health symptoms among Canadian adolescents. *Preventive Medicine*, 112, 168–75.

17. Dahl, R. E., Allen, N. B., Wilbrecht, L., & Suleiman, A. B. (2018). Importance of investing in adolescence from a developmental science perspective. *Nature*, 554(7693), 441–50; Sheehan, P., Sweeny, K., Rasmussen, B., et al. (2017). Building the foundations for sustainable

development: a case for global investment in the capabilities of adolescents. *The Lancet*, 390(10104), 1792–806.

18.   Price, E., Maguire, S., Firth, C., et al. (2022). Factors associated with nature connectedness in school-aged children. *Current Research in Ecological and Social Psychology*, 100037. https://doi.org/10.1016/j.cresp.2022.100037.

19.   FridaysForFuture. (2019). Fridaysforfuture.org. https://www.fridaysforfuture.org/

20.   Our demands. (n.d.). UK Student Climate Network. https://ukscn.org/our-demands/

21.   Fridays For Future – Inspiring climate activists. (n.d.). Fridays for Future. http://www.fridaysforfuture.org/greta-speeches

22.   Hughes, J., Richardson, M., & Lumber, R. (2018). Evaluating connection to nature and the relationship with conservation behaviour in children. *Journal for Nature Conservation*, 45, 11–19.

23.   Christis, M., Breemersch, K., Vercalsteren, A., & Dils, E. (2019). A detailed household carbon footprint analysis using expenditure accounts – case of Flanders (Belgium). *Journal of Cleaner Production*, 228, 1167–75.

24.   Soga, M., Yamanoi, T., Tsuchiya, K., et al. (2018). What are the drivers of and barriers to children's direct experiences of nature? *Landscape and Urban Planning*, 180, 114–20.

25.   Passmore, H. A., Martin, L., Richardson, M., et al. (2021). Parental/guardians' connection to nature better predicts children's nature connectedness than visits or area-level characteristics. *Ecopsychology*, 13(2), 103–13.

26.   Arbuthnott, K. D., Sutter, G. C., & Heidt, C. T. (2014). Natural history museums, parks, and connection with nature. *Museum Management and Curatorship*, 29(2), 102–21.

27.   Chang, C.-c., Cox, D. T. C., Fan, Q., et al. (2022). People's desire to be in nature and how they experience it are partially heritable.

28.   Hunt, A., Stewart, D., Burt, J., & Dillon, J. (2016). Monitor of Engagement with the Natural Environment: a pilot to develop an indicator of visits to the natural environment by children – results from years 1 and 2 (March 2013 to February 2015). *Natural England Commissioned Reports*, 208.

29.   Sutton, L. (2008). The state of play: disadvantage, play and children's well-being. *Social Policy and Society*, 7(4), 537–49.

30.   Lekies, K. S., & Brensinger, J. D. (2017). Childhood nature experiences across residential settings: rural, suburban, and urban. In C. Freeman, P. Tranter & T. Skelton (eds) *Risk, Protection, Provision and Policy* (Geographies of Children and Young People, Vol. 12) (pp. 67–86), Springer Verlag, Singapore.

31.   Lauricella, A. R., Cingel, D. P., Beaudoin-Ryan, L., et al. (2016). *The Common Sense Census: Plugged-In Parents of Tweens and Teens*. San Francisco, CA: Common Sense Media; Price, E., Maguire, S., Firth, C., et al. (2022). Factors associated with nature connectedness in school-aged children. *Current Research in Ecological and Social Psychology*, 3, 100037.

32. Mears, M., Brindley, P., Maheswaran, R., & Jorgensen, A. (2019). Understanding the socioeconomic equity of publicly accessible greenspace distribution: the example of Sheffield, UK. *Geoforum*, 103, 126–37; Leong, M., Dunn, R. R., & Trautwein, M. D. (2018). Biodiversity and socioeconomics in the city: a review of the luxury effect. *Biology Letters*, 14(5), 20180082.

33. Bakir-Demir, T., Berument, S. K., & Sahin-Acar, B. (2019). The relationship between greenery and self-regulation of children: the mediation role of nature connectedness. *Journal of Environmental Psychology*, 65, 101327.

34. Barrable, A. (2019). The case for nature connectedness as a distinct goal of early childhood education. *International Journal of Early Childhood Environmental Education*, 6(2), 59–70.

35. White, M. P., Elliott, L. R., Grellier, J. et al. (2021). Associations between green/blue spaces and mental health across 18 countries. *Scientific Reports*, 11, 8903. https://doi.org/10.1038/s41598-021-87675-0

36. Richardson, M., Hamlin, I., Elliott, L. R., & White, M. P. (2022). Country-level factors in a failing relationship with nature: nature connectedness as a key metric for a sustainable future. *Ambio*, 51, 1–13.

37. Eckersley, R. (2000). The mixed blessings of material progress: diminishing returns in the pursuit of happiness. *Journal of Happiness Studies*, 1(3), 267–92.

38. Kesebir, S., & Kesebir, P. (2017). A growing disconnection from nature is evident in cultural products. *Perspectives on Psychological Science*, 12(2), 258–69.

39. Richardson, M., Hussain, Z., & Griffiths, M. D. (2018). Problematic smartphone use, nature connectedness, and anxiety. *Journal of Behavioral Addictions*, 1–8.

40. Schmidt, L., & Marratto, S. L. (2008). *The End of Ethics in a Technological Society*. Montreal, QC: McGill-Queen's Press.

41. Lekies, K. S., & Brensinger, J. D. (2017). Childhood nature experiences across residential settings; Miller, J. R. (2005). Biodiversity conservation and the extinction of experience. *Trends in Ecology & Evolution*, 20(8), 430–4.

42. Nyborg, K., Anderies, J. M., Dannenberg, A., et al. (2016). Social norms as solutions. *Science*, 354(6308), 42–3.

43. Aminpour, P., Gray, S. A., Beck, M. W., et al. (2022). Urbanized knowledge syndrome – erosion of diversity and systems thinking in urbanites' mental models. *npj Urban Sustainability*, 2(1), 1–10.

44. Southon, G. E., Jorgensen, A., Dunnett, N., et al. (2017). Biodiverse perennial meadows have aesthetic value and increase residents' perceptions of site quality in urban green-space. *Landscape and Urban Planning*, 158, 105–18; Southon, G. E., Jorgensen, A., Dunnett, N., et al. (2018). Perceived species-richness in urban green spaces: cues, accuracy and well-being impacts. *Landscape and Urban Planning*, 172, 1–10; Cameron, R. W., Brindley, P., Mears, M., et al. (2020). Where the wild things are! Do urban green spaces with greater avian biodiversity promote more positive emotions in humans? *Urban Ecosystems*, 23(2), 301–17.

45. Pérez-Ramírez, I., García-Llorente, M., Saban de la Portilla, C., et al. (2021). Participatory collective farming as a leverage point for fostering human–nature connectedness. *Ecosystems and People*, 17(1), 222–34.

46. Egoz, S. (2000). Clean and green but messy: the contested landscape of New Zealand's organic farms. *Oral History*, 28(1), 63–74.

47. Balázsi, Á., Riechers, M., Hartel, T., et al. (2019). The impacts of social-ecological system change on human–nature connectedness: a case study from Transylvania, Romania. *Land Use Policy*, 89, 104232.

48. Pickett, K. E., & Wilkinson, R. G. (2015). Income inequality and health: a causal review. *Social Science & Medicine*, 128, 316–26.

Chapter 6: Good For You: Wellbeing Benefits of Reconnection

1. Rehm, J., & Shield, K. D. (2019). Global burden of disease and the impact of mental and addictive disorders. *Current Psychiatry Reports*, 21(2), 1–7.

2. Vigo, D., Thornicroft, G., & Atun, R. (2016). Estimating the true global burden of mental illness. *The Lancet Psychiatry*, 3(2), 171–8.

3. Nisbet, E. K., Zelenski, J. M., & Murphy, S. A. (2009). The nature relatedness scale: linking individuals' connection with nature to environmental concern and behavior. *Environment and Behavior*, 41(5), 715–40.

4. Zelenski, J. M., & Nisbet, E. K. (2014). Happiness and feeling connected: the distinct role of nature relatedness. *Environment and Behavior*, 46(1), 3–23.

5. Cartwright, B. D., White, M. P., & Clitherow, T. J. (2018). Nearby nature 'buffers' the effect of low social connectedness on adult subjective wellbeing over the last 7 days. *International Journal of Environmental Research and Public Health*, 15(6), 1238.

6. Howell, A. J., Passmore, H. A., & Buro, K. (2013). Meaning in nature: meaning in life as a mediator of the relationship between nature connectedness and well-being. *Journal of Happiness Studies*, 14(6), 1681–96.

7. Cervinka, R., Röderer, K., & Hefler, E. (2012). Are nature lovers happy? On various indicators of well-being and connectedness with nature. *Journal of Health Psychology*, 17(3), 379–88.

8. Nisbet, E. K., Zelenski, J. M., & Murphy, S. A. (2009). The nature relatedness scale; Di Fabio, A., & Rosen, M. A. (2019). Accounting for individual differences in connectedness to nature: personality and gender differences. *Sustainability*, 11(6), 1693; Lee, K., Ashton, M. C., Choi, J., & Zachariassen, K. (2015). Connectedness to nature and to humanity: their association and personality correlates. *Frontiers in Psychology*, 6, 1003.

9. Richardson, M., & Sheffield, D. (2015). Reflective self-attention: a more stable predictor of connection to nature than mindful attention. *Ecopsychology*, 7(3), 166–75.

10. Brown, K. W., & Ryan, R. M. (2003). The benefits of being present: mindfulness and its role in psychological well-being. *Journal of Personality and Social Psychology*, 84(4), 822.

11. Howell, A. J., Dopko, R. L., Passmore, H. A., & Buro, K. (2011). Nature connectedness: associations with well-being and mindfulness. *Personality and Individual Differences*, 51(2), 166–71.

12. Nisbet, E. K., Zelenski, J. M., & Grandpierre, Z. (2019). Mindfulness in nature enhances connectedness and mood. *Ecopsychology*, 11(2), 81–91.

13. Swami, V., Barron, D., Weis, L., & Furnham, A. (2016). Bodies in nature: associations between exposure to nature, connectedness to nature, and body image in US adults. *Body Image*, 18, 153–61; Holloway, J. A., Murray, J., Okada, R., & Emmons, A. L. (2014). Ecopsychology and relationship competency: the empowerment of women graduate students through nature experiences. *Women & Therapy*, 37(1–2), 141–54.

14. Fido, D., Rees, A., Clarke, P., et al. (2020). Examining the connection between nature connectedness and dark personality. *Journal of Environmental Psychology*, 72, 101499; Fido, D., & Richardson, M. (2019). Empathy mediates the relationship between nature connectedness and both callous and uncaring traits. *Ecopsychology*, 11(2).

15. Leong, L. Y. C., Fischer, R., & McClure, J. (2014). Are nature lovers more innovative? The relationship between connectedness with nature and cognitive styles. *Journal of Environmental Psychology*, 40, 57–63.

16. Pritchard, A., Richardson, M., Sheffield, D., & McEwan, K. (2020). The relationship between nature connectedness and eudaimonic well-being: a meta-analysis. *Journal of Happiness Studies*, 21(3), 1145–67.

17. Ridder, B. (2005). Reorienting environmentalism to nature-inspired-autonomy. *Griffith Journal of the Environment*, 1, 1–26.

18. Jefferies, R. (1884). *Life of the Fields*. London: Chatto and Windus.

19. Capra, F. (1983). *The Turning Point: Science, Society, and the Rising Culture*. New York: Bantam; Lent, J. R. (2017). *The Patterning Instinct: A Cultural History of Humanity's Search for Meaning* (p. 409). Amherst, NY: Prometheus Books.

20. Martin, L., White, M. P., Hunt, A., et al. (2020). Nature contact, nature connectedness and associations with health, wellbeing and pro-environmental behaviours. *Journal of Environmental Psychology*, 68, 101389.

21. Richardson, M., Passmore, H. A., Lumber, R., et al. (2021). Moments, not minutes: The nature–wellbeing relationship. *International Journal of Wellbeing*, 11(1), 8–33.

22. Kraha, A., Turner, H., Nimon, K., et al. (2012). Tools to support interpreting multiple regression in the face of multicollinearity. *Frontiers in Psychology*, 3, 44.

23. Liu, H., Nong, H., Ren, H., & Liu, K. (2022). The effect of nature exposure, nature connectedness on mental well-being and ill-being in a general Chinese population. *Landscape and Urban Planning*, 222, 104397.

24.   Richardson, M. & Hamlin, I. (2021) Nature engagement for human and nature's Wellbeing during the Corona pandemic. *Journal of Public Mental Health*. https://doi.org/10.1108/JPMH-02-2021-0016

25.   Tester-Jones, M., White, M. P., Elliott, L. R., et al. (2020). Results from an 18 country cross-sectional study examining experiences of nature for people with common mental health disorders. *Scientific Reports*, 10(1), 1–11. https://doi.org/10.1038/s41598-020-75825-9

Chapter 7: How Does Reconnection Bring Wellbeing?

1.   Brymer, E., Cuddihy, T. F., & Sharma-Brymer, V. (2010). The role of nature-based experiences in the development and maintenance of wellness. *Asia-Pacific Journal of Health, Sport and Physical Education*, 1(2), 21–7; Wilson, E. O. (1984). *Biophilia*. Cambridge, MA: Harvard University Press; Davis, J. (1998). The transpersonal dimensions of ecopsychology: nature, nonduality, and spiritual practice. *The Humanistic Psychologist*, 26(1–3), 69-100.

2.   Segal, F. (1997). Ecopsychology and the uses of wilderness. *Ecopsychology On-Line*, 5; Duncan, G. (1998). The psychological benefits of wilderness. *Ecopsychology On-Line*, 6.

3.   Pilisuk, M. (2001). Ecological psychology, caring, and the boundaries of the person. *Journal of Humanistic Psychology*, 41(2), 25–37.

4.   Roszak, T. (1995). Where psyche meets Gaia. In T. Roszak, M. E. Gomes & A. D. Kanner (eds) *Ecopsychology: Restoring the Earth, Healing the Mind* (pp. 1–17). San Francisco: Sierra Club Books; Roszak, T. (1992). *The Voice of the Earth*. New York: Touchstone.

5.   Fabrega Jr, H. (2004). Psychiatric conditions in an evolutionary context. *Psychopathology*, 37(6), 290–8; Fabrega, H. (2002). *Origins of Psychopathology: The Phylogenetic and Cultural Basis of Mental Illness*. New Brunswick, NJ: Rutgers University Press.

6.   Brüne, M., Brüne-Cohrs, U., McGrew, W. C., & Preuschoft, S. (2006). Psychopathology in great apes: concepts, treatment options and possible homologies to human psychiatric disorders. *Neuroscience & Biobehavioral Reviews*, 30(8), 1246–59.

7.   Birkett, L. P., & Newton-Fisher, N. E. (2011). How abnormal is the behaviour of captive, zoo-living chimpanzees? *PLoS ONE*, 6(6), e20101.

8.   Clarke, A. S., Juno, C. J., & Maple, T. L. (1982). Behavioral effects of a change in the physical environment: a pilot study of captive chimpanzees. *Zoo Biology*, 1(4), 371–80.

9.   Goodall, J. (1986). *The Chimpanzees of the Gombe: Patterns of Behavior* (p. 673). Boston, MA: Harvard University Press; Brüne, M., Brüne-Cohrs, U., McGrew, W. C., & Preuschoft, S. (2006) Psychopathology in great apes: concepts, treatment options and possible homologies to human psychiatric disorders. *Neuroscience & Biobehavioral Reviews*, 30(8), 1246–59.

10.   Ferdowsian, H. R., Durham, D. L., Kimwele, C., et al. (2011). Signs of mood and anxiety disorders in chimpanzees. *PloS ONE*, 6(6), e19855;

Baker, C. (2021, December 13). Mental health statistics for England: prevalence, services and funding. Parliament.uk. https://researchbriefings.files.parliament.uk/documents/SN06988/SN06988.pdf

11.   McMahan, E. A., & Estes, D. (2015). The effect of contact with natural environments on positive and negative affect: a meta-analysis. *The Journal of Positive Psychology*, 10(6), 507–19.

12.   Hartig, T., van den Berg, A. E., Hagerhall, C. M., et al. (2011). Health benefits of nature experience: psychological, social and cultural processes. In K. Nilsson, M. Sangster, C. Gallis, et al. (eds) *Forests, Trees and Human Health* (pp. 127–68). Dordrecht: Springer.

13.   Kaplan, S. (1995). The restorative benefits of nature: toward an integrative framework. *Journal of Environmental Psychology*, 15(3), 169–82; Ulrich, R. S., Simons, R. F., Losito, B. D., et al. (1991). Stress recovery during exposure to natural and urban environments. *Journal of Environmental Psychology*, 11(3), 201–30.

14.   Gidlow, C. J., Jones, M. V., Hurst, G., et al. (2016). Where to put your best foot forward: psycho-physiological responses to walking in natural and urban environments. *Journal of Environmental Psychology*, 45, 22–9. https://doi.org/10.1016/j.jenvp.2015.11.003; Capaldi, C. A., Passmore, H. A., Ishii, R., et al. (2017). Engaging with natural beauty may be related to well-being because it connects people to nature: evidence from three cultures. *Ecopsychology*, 9, 199–211. https://doi.org/10.1089/eco.2017.0008

15.   Beute, F., & de Kort, Y. A. (2014). Salutogenic effects of the environment: review of health protective effects of nature and daylight. *Applied Psychology: Health and Well-Being*, 6(1), 67–95.

16.   Johnsen, S. Å. K., & Rydstedt, L. W. (2013). Active use of the natural environment for emotion regulation. *Europe's Journal of Psychology*, 9(4), 798–819.

17.   Zhang, J. W., Howell, R. T., & Iyer, R. (2014). Engagement with natural beauty moderates the positive relation between connectedness with nature and psychological well-being. *Journal of Environmental Psychology*, 38, 55–63.

18.   Capaldi, C. A., Passmore, H. A., Ishii, R., et al. (2017). Engaging with natural beauty may be related to well-being because it connects people to nature: evidence from three cultures. *Ecopsychology* 9, 199–211. https://doi.org/10.1089/eco.2017.0008

19.   Richardson, M., & McEwan, K. (2018). 30 Days Wild and the relationships between engagement with nature's beauty, nature connectedness and well-being. *Frontiers in Psychology*, 9, 1500.

20.   Bakir-Demir, T., Berument, S. K., & Akkaya, S. (2021). Nature connectedness boosts the bright side of emotion regulation, which in turn reduces stress. *Journal of Environmental Psychology*, 101642.

21.   Reber, R., Schwarz, N., & Winkielman, P. (2004). Processing fluency and aesthetic pleasure: is beauty in the perceiver's processing experience? *Personality and Social Psychology Review*, 8, 364–82. https://doi.

org/10.1207/s15327957pspr0804_3; Kaplan, S. (1987). Aesthetics, affect, and cognition: environmental preference from an evolutionary perspective. *Environment and Behavior*, 19(1), 3–32.

22.   Ryan, R. M., & Deci, E. L. (2001). On happiness and human potentials: a review of research on hedonic and eudaimonic well-being. *Annual Review of Psychology*, 52(1), 141–66; Cleary, A., Fielding, K. S., Bell, S. L., et al. (2017). Exploring potential mechanisms involved in the relationship between eudaimonic wellbeing and nature connection. *Landscape and Urban Planning*, 158, 119–28; Nisbet, E. K., Zelenski, J. M., & Murphy, S. A. (2009). The nature relatedness scale: linking individuals' connection with nature to environmental concern and behavior. *Environment and Behavior*, 41, 715–40; Howell, A. J., & Passmore, H. (2013). The nature of happiness: nature affiliation and mental well-being. In C. L. M. Keyes (ed.) *Mental Well-Being: International Contributions to the Study of Positive Mental Health* (pp. 231–57). New York: Springer; Pritchard, A., Richardson, M., Sheffield, D., & McEwan, K. (2020). The relationship between nature connectedness and eudaimonic well-being: a meta-analysis. *Journal of Happiness Studies*, 21(3), 1145–67.

23.   Kamitsis, I., & Francis, A. J. (2013). Spirituality mediates the relationship between engagement with nature and psychological wellbeing. *Journal of Environmental Psychology*, 36, 136–43; Trigwell, J. L., Francis, A. J., & Bagot, K. L. (2014). Nature connectedness and eudaimonic well-being: spirituality as a potential mediator. *Ecopsychology*, 6(4), 241–51.

24.   Howell, A. J., Passmore, H. A., & Buro, K. (2013). Meaning in nature: meaning in life as a mediator of the relationship between nature connectedness and well-being. *Journal of Happiness Studies*, 14(6), 1681–96.

25.   Passmore, H. A., & Howell, A. J. (2014). Eco-existential positive psychology: experiences in nature, existential anxieties, and well-being. *The Humanistic Psychologist*, 42(4), 370–88.

26.   Aruta, J. J. B. R. (2021). The quest to mental well-being: nature connectedness, materialism and the mediating role of meaning in life in the Philippine context. *Current Psychology*, 1–12.

27.   Maohong, B. (2012). Deforestation in the Philippines, 1946–1995. *Philippine Studies: Historical and Ethnographic Viewpoints*, 60(1), 117–30. https://doi.org/10.1353/phs.2012.0011; Gultiano, S., & Xenos, P. (2004). Age-structure and urban migration of youth in the Philippines. In *Paper Presented at the CICRED Seminar Age-Structured Transitions: Demographic Bonuses, but Emerging Challenges for Population and Sustainable Development*, 23–26 February, Paris; Malaque III, I. R., & Yokohari, M. (2007). Urbanization process and the changing agricultural landscape pattern in the urban fringe of metro Manila, Philippines. *Environment & Urbanization*, 19(1), 191–206. https://doi.org/10.1177/0956247807076782

28.   Philippines. (2020). Sea Circular. https://www.sea-circular.org/country/philippines/

Chapter 8: Good for Nature: Environmental Benefits of Reconnection

1.   Mackay, C. M., & Schmitt, M. T. (2019). Do people who feel connected to nature do more to protect it? A meta-analysis. *Journal of Environmental Psychology*, 65, 101323.

2.   Whitburn, J., Linklater, W., & Abrahamse, W. (2020). Meta-analysis of human connection to nature and proenvironmental behavior. *Conservation Biology*, 34(1), 180–93.

3.   New UNEP synthesis provides blueprint to urgently solve planetary emergencies and secure humanity's future. (2021). UN Environment. https://www.unep.org/news-and-stories/press-release/new-unep-synthesis-provides-blueprint-urgently-solve-planetary

4.   Richardson, M., Hunt, A., Hinds, J., et al. (2019). A measure of nature connectedness for children and adults: validation, performance, and insights. *Sustainability*, 11(12), 3250. https://doi.org/10.3390/su11123250

5.   Hughes, J., Richardson, M., & Lumber, R. (2018). Evaluating connection to nature and the relationship with conservation behaviour in children. *Journal for Nature Conservation*, 45, 11–19.

6.   Legagneux, P., Casajus, N., Cazelles, K., et al. (2018). Our house is burning: discrepancy in climate change vs. biodiversity coverage in the media as compared to scientific literature. *Frontiers in Ecology and Evolution*, 5, 175.

7.   Hughes, J., Richardson, M., & Lumber, R. (2018). Evaluating connection to nature.

8.   Martin, L., White, M. P., Hunt, A., et al. (2020). Nature contact, nature connectedness and associations with health, wellbeing and pro-environmental behaviours. *Journal of Environmental Psychology*, 68, 101389.

9.   Jongman, R. H. (1995). Nature conservation planning in Europe: developing ecological networks. *Landscape and Urban Planning*, 32(3), 169–83.

10. Goddard, M. A., Dougill, A. J., & Benton, T. G. (2010). Scaling up from gardens: biodiversity conservation in urban environments. *Trends in Ecology & Evolution*, 25(2), 90–8.

11. Telesetsky, A. (2017). Eco-restoration, private landowners and overcoming the status quo bias. *Griffith Law Review*, 26(2), 248–74; van Heezik, Y. M., Dickinson, K. J., & Freeman, C. (2012). Closing the gap: communicating to change gardening practices in support of native biodiversity in urban private gardens. *Ecology and Society*, 17(1), 34.

12. Gooden, J. L. (2019). Cultivating identity through private land conservation. *People and Nature*, 1(3), 362–75.

13. Barbett, L., Stupple, E., Sweet, M., & Richardson, M. (2019, November 5). *An Expert Ranked List of Pro-Nature Conservation Behaviours for Public Use*. https://doi.org/10.31234/osf.io/bzmsv

14. Van Heezik, Y., Freeman, C., Davidson, K., & Lewis, B. (2020). Uptake and engagement of activities to promote native species in private gardens. *Environmental Management*, 66(1), 42–55.

15. Barbett, L., Stupple, E. J., Sweet, M., Schofield, M. B., & Richardson, M. (2020). Measuring actions for nature – development and validation of a pro-nature conservation behaviour scale. *Sustainability*, 12(12), 4885.

16. Rose, L. (2020). *Framing Nature: Conservation and Culture.* Hebden Bridge: Gritstone Publishing Co-operative.

Chapter 9: One Health

1. Richardson, M., Passmore, H. A., Barbett, L., et al. (2020). The green care code: how nature connectedness and simple activities help explain pro-nature conservation behaviours. *People and Nature*, 2(3), 821–39.

2. BBC (2019). The state of nature, *Costing the Earth*. www.bbc.co.uk/programmes/m0004sj9

3. Sato, M., Aoshima, I., & Chang, Y. (2021). Connectedness to nature and the conservation of the urban ecosystem: perspectives from the valuation of urban forests. *Forest Policy and Economics*, 125, 102396.

4. Knapp, J. L., Phillips, B. B., Clements, J., et al. (2021). Socio-psychological factors, beyond knowledge, predict people's engagement in pollinator conservation. *People and Nature*, 3(1), 204–20.

5. Sturm, U., Straka, T. M., Moormann, A., & Egerer, M. (2021). Fascination and joy: emotions predict urban gardeners' pro-pollinator behaviour. *Insects*, 12(9), 785.

6. Richardson, M. & Hamlin, I. (2021) Nature engagement for human and nature's wellbeing during the Corona pandemic. *Journal of Public Mental Health*. https://doi.org/10.1108/JPMH-02-2021-0016

7. Hamlin, I. & Richardson, M. (2021). *Visible Garden Biodiversity Leads to an Increase in Noticing Nature, Which in Turn Leads to an Increase in Nature Connectedness.* https://doi.org/10.31234/osf.io/uamwg

8. Blake, J. (1999). Overcoming the 'value-action gap' in environmental policy: tensions between national policy and local experience. *Local Environment*, 4(3), 257–78.

9. Russell, S. V., Young, C. W., Unsworth, K. L., & Robinson, C. (2017). Bringing habits and emotions into food waste behaviour. *Resources, Conservation and Recycling*, 125, 107–14; Bamberg, S., & Möser, G. (2007). Twenty years after Hines, Hungerford, and Tomera: a new meta-analysis of psycho-social determinants of pro-environmental behaviour. *Journal of Environmental Psychology*, 27(1), 14–25; Weiss, H. M., & Beal, D. J. (2005). Reflections on affective events theory. In N. Ashkanasy, C. Härtel & W. J. Zerbe (eds) *The Effect of Affect in Organizational Settings*. Bingley, UK: Emerald Group Publishing.

10. Barbett, L. (2021). What can we do for nature? A systematic research approach to pro-nature conservation behaviours. PhD thesis, University of Derby.

11. Barragan-Jason, G., de Mazancourt, C., Parmesan, C., et al. (2021). Human–nature connectedness as a pathway to sustainability: a global meta-analysis. *Conservation Letters*, e1285.

12. *The Sun*. (2017). Millions of Brits are clueless about feathered friends and half can't id SPARROW. https://www.thesun.co.uk/news/3272662/millions-of-brits-unable-to-identify-sparrow/

13. Stroud Greens. (n.d.). Twitter. https://twitter.com/StroudGreens/status/1201604310838267909?s=20

14. Louv, R. (2012). *The Nature Principle: Reconnecting with Life in a Virtual Age*. Chapel Hill, NC: Algonquin Books.

15. Simpson, J. (2008). *The Oxford Pocket Dictionary of Current English*. Oxford: Oxford University Press.

16. Bird-David, N. (1999). 'Animism' revisited: personhood, environment, and relational epistemology. *Current Anthropology*, 40(S1), S67–91.

17. Niigaaniin, M., & MacNeill, T. (2022). Indigenous culture and nature relatedness: Results from a collaborative study. *Environmental Development*, 44, 100753.

18. Shepherd, N. (2008). *The Living Mountain: A Celebration of the Cairngorm Mountains of Scotland*. Edinburgh: Canongate Books.

19. Otto, S., & Pensini, P. (2017). Nature-based environmental education of children: environmental knowledge and connectedness to nature, together, are related to ecological behaviour. *Global Environmental Change*, 47, 88–94.

20. Barragan-Jason, G., de Mazancourt, C., Parmesan, C., et al. (2021). Human–nature connectedness as a pathway to sustainability: a global meta-analysis. *Conservation Letters*, e1285.

21. Bettencourt, L. M., Lobo, J., Helbing, D., et al. (2007). Growth, innovation, scaling, and the pace of life in cities. *Proceedings of the National Academy of Sciences*, 104(17), 7301–6.

22. Hoffmann, C., Hoppe, J. A., & Ziemann, N. (2022). Faster, harder, greener? Empirical evidence on the role of the individual Pace of Life for productivity and pro-environmental behavior. *Ecological Economics*, 191, 107212; Aruta, J. J. B. R. (2021). Connectedness to nature encourages, but materialism hinders, ecological behavior in the Philippines: the higher order and second-order factors of environmental attitudes as viable mediating pathways. *Ecopsychology*, 13(2), 114–22.

23. Barrows, P. D., Richardson, M., Hamlin, I., & Van Gordon, W. (2022). Nature connectedness, nonattachment, and engagement with nature's beauty predict pro-nature conservation behavior. *Ecopsychology*, 14(2), 83–91.

24. Schultz, P., & Kaiser, F. G. (2012). Promoting pro-environmental behavior. In S. D. Clayton (ed.) *The Oxford Handbook of Environmental and Conservation Psychology* 556-580. Oxford: Oxford University Press.

25. Gustafson, A., Pace, A., Singh, S., & Goldberg, M. (2022). What do people say is the most important reason to protect nature? An analysis of pro-environmental motives across 11 countries. *Journal of Environmental Psychology*, 80, 101762.

26. Rose, L. (2021). *Leopard Moon Rising: Distant Views of India*. Hebden Bridge: Gritstone Publishing Cooperative.

27.   DeMello, M. (2021). The animal protection movement. In *Animals and Society* (pp. 470–501). New York: Columbia University Press; Lavigne, D., Kidman Cox, R., Menon, V., & Wamithi, M. (2006). Reinventing wildlife conservation for the 21st century. In D. M. Lavigne (ed.) *Gaining Ground: In Pursuit of Ecological Sustainability* (pp. 379–406). London, ON: International Fund for Animal Welfare, Guelph, Canada, and the University of Limerick; Rich, B. (2010). *To Uphold the World: A Call for a New Global Ethic from Ancient India*. Boston: Beacon Press.

28.   Dhand, A. (2002). The dharma of ethics, the ethics of dharma: quizzing the ideals of Hinduism. *Journal of Religious Ethics*, 30(3), 347–72.

29.   Jain, P. (2016). *Dharma and Ecology of Hindu Communities: Sustenance and Sustainability*. London: Routledge.

30.   Alam, K., & Halder, U. K. (2018). A pioneer of environmental movements in India: Bishnoi movement. *Journal of Education and Development*, 8(15), 283–7.

31.   Khoshoo, T. N. (1999). The dharma of ecology. *Current Science*, 77(9), 1147–53.

32.   Rabinowitz, P. M., Pappaioanou, M., Bardosh, K. L., & Conti, L. (2018). A planetary vision for one health. *BMJ Global Health*, 3(5), e001137; Nelson, D. H., Prescott, S. L., Logan, A. C., & Bland, J. S. (2019). Clinical ecology – transforming 21st-century medicine with planetary health in mind. *Challenges*, 10(1), 15; Lerner, H., & Berg, C. (2017). A comparison of three holistic approaches to health: One Health, EcoHealth, and Planetary Health. Frontiers in Veterinary Science, 4, 163; Brymer, E., Freeman, E., & Richardson, M. (2019). One health: the well-being impacts of human–nature relationships. *Frontiers in Psychology*, 10, 1611.

33.   Richardson, M., Maspero, M., Golightly, D., et al. (2017). Nature: a new paradigm for well-being and ergonomics. *Ergonomics*, 60(2), 292–305.

34.   Lerner, H., & Berg, C. (2017). A comparison of three holistic approaches to health.

35.   Stevens, P. (2010). Embedment in the environment: a new paradigm for well-being? *Perspectives in Public Health*, 130(6), 265–9.

36.   Edmonds, D. (2001). *Electricity and Magnetism in Biological Systems*. Oxford: Oxford University Press; Stevens, P. (2007). Affective response to 5 µT ELF magnetic field-induced physiological changes. *Bioelectromagnetics: Journal of the Bioelectromagnetics Society, the Society for Physical Regulation in Biology and Medicine, the European Bioelectromagnetics Association*, 28(2), 109–14; Cook, C. M., Saucier, D. M., Thomas, A. W., & Prato, F. S. (2006). Exposure to ELF magnetic and ELF-modulated radiofrequency fields: the time course of physiological and cognitive effects observed in recent studies (2001–2005). *Bioelectromagnetics: Journal of the Bioelectromagnetics Society, the Society for Physical Regulation in Biology and Medicine, the European Bioelectromagnetics Association*, 27(8), 613–27; Mutlu, G. M., Garey, K. W., Robbins, R. A., et al. (2001). Collection and analysis of exhaled breath

condensate in humans. *American Journal of Respiratory and Critical Care Medicine*, 164(5), 731–7.

37. Making Peace with Nature. (n.d.). https://wedocs.unep.org/xmlui/bitstream/handle/20.500.11822/34948/MPN.pdf

Chapter 10: The Good Things in Nature

1. Barton, J., Bragg, R., Pretty, J., et al. (2016). The wilderness expedition: an effective life course intervention to improve young people's well-being and connectedness to nature. *Journal of Experiential Education*, 39(1), 59–72.

2. Frantz, C. M., & Mayer, F. S. (2014). The importance of connection to nature in assessing environmental education programs. *Studies in Educational Evaluation*, 41, 85–9. https://doi.org/10.1016/j.stueduc.2013.10.001; Mayer, F. S., Frantz, C. M., Bruehlman-Senecal, E., & Dolliver, K. (2009). Why is nature beneficial? The role of connectedness to nature. *Environment and Behavior*, 41, 607–43. https://doi.org/10.1177/0013916508319745; Nisbet, E. K., & Zelenski, J. M. (2013). The NR-6: a new brief measure of nature relatedness. *Frontiers in Psychology*, 4, 813. https://doi.org/10.3389/fpsyg.2013.00813

3. Seligman, M. E., Steen, T. A., Park, N., & Peterson, C. (2005). Positive psychology progress: empirical validation of interventions. *American Psychologist*, 60(5), 410; Emmons, R. A., & McCullough, M. E. (2003). Counting blessings versus burdens: an experimental investigation of gratitude and subjective well-being in daily life. *Journal of Personality and Social Psychology*, 84(2), 377–89.

4. Richardson, M., & Sheffield, D. (2019). *The Negative Impact of a Three Good Things Intervention on Perceived Stress and Psychological Health*. https://doi.org/10.31234/osf.io/p463y

5. Richardson, M., & Hallam, J. (2013). Exploring the psychological rewards of a familiar semirural landscape: connecting to local nature through a mindful approach. *The Humanistic Psychologist*, 41(1), 35.

6. Richardson, M., & Sheffield, D. (2017). Three good things in nature: noticing nearby nature brings sustained increases in connection with nature. *PsyEcology*, 8(1), 1–32.

7. Burton, C. M., & King, L. A. (2008). Effects of (very) brief writing on health: the two-minute miracle. *British Journal of Health Psychology*, 13(1), 9–14; Menary, R. (2007). Writing as thinking. *Language Sciences*, 29(5), 621–32.

8. Clark, A., & Chalmers, D. (1998). The extended mind. *Analysis*, 58(1), 7–19; Thompson, E. (2010). *Mind in Life*. Cambridge, MA: Harvard University Press.

9. Merleau-Ponty, M. (1968). *The Visible and the Invisible: Followed by Working Notes*. Evanston, IL: Northwestern University Press; Schroeder, H. W. (1991). Preference and meaning of arboretum landscapes: combining quantitative and qualitative data. *Journal of Environmental Psychology*, 11(3), 231–48.

10.   Sabloff, A., & Lemon, J. (2001). Reordering the natural world: humans & animals in the city. *Urban History Review*, 30(1), 71.

11.   Clark, A., (1997). *Being There*. Cambridge, MA: MIT Press; Gallagher, S. (2005). *How the Body Shapes the Mind*. Oxford: Oxford University Press; Lakoff, G., & Johnson, M. (1999). *Philosophy in the Flesh: The Embodied Mind and its Challenge to Western Thought*. New York: Basic Books; Jacob, P. (2012). Embodying the mind by extending it. *Review of Philosophy and Psychology*, 3(1), 33–51; Noë, A. (2009). *Out of Our Heads: Why You Are Not Your Brain, and Other Lessons from the Biology of Consciousness*. New York: Macmillan.

12.   McEwan, K., Richardson, M., Brindley, P., et al. (2020). Shmapped: development of an app to record and promote the well-being benefits of noticing urban nature. *Translational Behavioral Medicine*, 10(3), 723–33; McEwan, K., Richardson, M., Sheffield, D., et al. (2021). Assessing the feasibility of public engagement in a smartphone app to improve well-being through nature connection. *PsyEcology*, 12(1), 45–75.

13.   McEwan, K., Richardson, M., Sheffield, D., et al. (2019). A smartphone app for improving mental health through connecting with urban nature. *International Journal of Environmental Research and Public Health*, 16(18), 3373.

14.   Gilbert, P., McEwan, K., Mitra, R., et al. (2008). Feeling safe and content: a specific affect regulation system? Relationship to depression, anxiety, stress, and self-criticism. *Journal of Positive Psychology*, 3(3), 182–91.

15.   McManus, M. D., Siegel, J. T., & Nakamura, J. (2019). The predictive power of low-arousal positive affect. *Motivation and Emotion*, 43(1), 130–44.

16.   Nature Prescriptions Supporting the health of people and nature. (n.d.). https://www.rspb.org.uk/globalassets/downloads/documents/nature-prescriptions/Edinburgh-pilot-final-report.pdf; RSPB. (n.d.). November. https://community.rspb.org.uk/cfs-file/__key/telligent-evolution-components-attachments/01-52681-00-00-00-79-16-81/RSPB-Nature-Prescriptions-November-Calendar.pdf; Nature prescriptions helping hundreds of patients in Edinburgh. (n.d.). The RSPB. https://www.rspb.org.uk/about-the-rspb/about-us/media-centre/press-releases/nature-prescriptions-edinburgh-results/

17.   Sturm, V. E., Datta, S., Roy, A. R. K., et al. (2020). Big smile, small self: awe walks promote prosocial positive emotions in older adults. *Emotion*. Advance online publication. https://doi.org/10.1037/emo0000876

18.   Keenan, R., Lumber, R., Richardson, M. and Sheffield, D. (2021), Three good things in nature: a nature-based positive psychological intervention to improve mood and well-being for depression and anxiety. *Journal of Public Mental Health*. Advance online publication. https://doi.org/10.1108/JPMH-02-2021-0029

19.   Bakker, D., Kazantzis, N., Rickwood, D., & Rickard, N. (2018). A randomized controlled trial of three smartphone apps for enhancing public mental health. *Behaviour Research and Therapy*, 109, 75–83.

20.   Kawadler, J. M., Hemmings, N. R., Ponzo, S., et al. (2020). Effectiveness of a smartphone app (BioBase) for reducing anxiety and increasing

mental well-being: pilot feasibility and acceptability study. *JMIR Formative Research*, 4(11), e18067; Powell, J., Hamborg, T., Stallard, N., et al. (2013). Effectiveness of a web-based cognitive-behavioral tool to improve mental well-being in the general population: randomized controlled trial. *Journal of Medical Internet Research*, 15(1), e2; Bhutani, G. E. (2015). Looking after me looking after you: using positive cognitive behavioural techniques to improve emotional well-being. *The Cognitive Behaviour Therapist*, 8.

21.   Beshai, S., McAlpine, L., Weare, K., & Kuyken, W. (2016). A non-randomised feasibility trial assessing the efficacy of a mindfulness-based intervention for teachers to reduce stress and improve well-being. *Mindfulness*, 7(1), 198–208.

22.   Maund, P. R., Irvine, K. N., Reeves, J., et al. (2019). Wetlands for wellbeing: piloting a nature-based health intervention for the management of anxiety and depression. *International Journal of Environmental Research and Public Health*, 16(22), 4413.

23.   Pocock, M., Hamlin, I., Christelow, J., et al. (In press). The benefits of citizen science and nature-noticing activities for wellbeing, nature connectedness and pro-nature conservation behaviours. People & Nature.

24.   Richardson, M., Hallam, J., & Lumber, R. (2015). One thousand good things in nature: aspects of nearby nature associated with improved connection to nature. *Environmental Values*, 24(5), 603–19.

25.   McEwan, K., Ferguson, F. J., Richardson, M., & Cameron, R. (2020). The good things in urban nature: a thematic framework for optimising urban planning for nature connectedness. *Landscape and Urban Planning*, 194, 103687.

26.   Investigation into Sheffield tree felling scandal still ongoing due to "volume of documents involved." (2021). www.yorkshirepost. co.uk. https://www.yorkshirepost.co.uk/news/politics/investigation-into-sheffield-tree-felling-scandal-still-ongoing-due-to-volume-of-documents-involved-3355297

Chapter 11: Pathways to Reconnection

1.   Richardson, M., Hamlin, I., Butler, C. W., et al. (2021). Actively noticing nature (not just time in nature) helps promote nature connectedness. *Ecopsychology*. https://doi.org/10.1089/eco.2021.0023

2.   Colléony, A., Levontin, L., & Shwartz, A. (2020). Promoting meaningful and positive nature interactions for visitors to green spaces. *Conservation Biology*, 34(6), 1373–82.

3.   Passmore, H. A., & Holder, M. D. (2017). Noticing nature: individual and social benefits of a two-week intervention. *Journal of Positive Psychology*, 12(6), 537–46.

4.   Passmore, H. A., Yang, Y., & Sabine, S. (2022). An extended replication study of the well-being intervention, the noticing nature intervention (NNI). *Journal of Happiness Studies*, 1–21.

5. Barrable, A., & Booth, D. (2020). Green and screen: does mobile photography enhance or hinder our connection to nature? *Digital Culture & Education*, 12(2).

6. Bishop, S. R., Lau, M., Shapiro, S., et al. (2004). Mindfulness: a proposed operational definition. *Clinical Psychology: Science and Practice*, 11(3), 230.

7. Barrable, A., Booth, D., Adams, D., & Beauchamp, G. (2021). Enhancing nature connection and positive affect in children through mindful engagement with natural environments. *International Journal of Environmental Research and Public Health*, 18(9), 4785.

8. Bingjing, C., Chen, G., & Shuhua, L. I. (2022). Looking at buildings or trees? Association of human nature relatedness with eye movements in outdoor space. *Journal of Environmental Psychology*, 101756.

9. Douglas, J. W., & Evans, K. L. (2021). An experimental test of the impact of avian diversity on attentional benefits and enjoyment of people experiencing urban green-space. *People and Nature*. https://doi.org/10.1002/pan3.10279

10. Batool, A., Rutherford, P., McGraw, P., et al. (2021). Gaze correlates of view preference: comparing natural and urban scenes. *Lighting Research & Technology*, https://doi.org/10.1177/14771535211055703

11. Sheffield, D., Butler, C. W., & Richardson, M. (2022). Improving Nature Connectedness in Adults: A Meta-Analysis, Review and Agenda. *Sustainability*, 14(19), 12494.

12. Pharo, C. (2017). Shock as millions of bird-brained Brits can't spot a common sparrow. *Mirror*. https://www.mirror.co.uk/news/uk-news/millions-bird-brained-brits-cant-10173651

13. Mayer, F. S., Frantz, C. M., Bruehlman-Senecal, E., & Dolliver, K. (2009). Why is nature beneficial? The role of connectedness to nature. *Environment and Behavior*, 41(5), 607–43; Schultz, P. W., & Tabanico, J. (2007). Self, identity, and the natural environment: exploring implicit connections with nature 1. *Journal of Applied Social Psychology*, 37(6), 1219–47; Arbuthnott, K. D., Sutter, G. C., & Heidt, C. T. (2014). Natural history museums, parks, and connection with nature. *Museum Management and Curatorship*, 29(2), 102–21.

14. Liu, H., Nong, H., Ren, H., & Liu, K. (2022). The effect of nature exposure, nature connectedness on mental well-being and ill-being in a general Chinese population. *Landscape and Urban Planning*, 222, 104397.

15. Williams, I. R., Rose, L. M., Raniti, M. B., et al. (2018). The impact of an outdoor adventure program on positive adolescent development: a controlled crossover trial. *Journal of Outdoor and Environmental Education*, 21(2), 207–36.

16. Martin, P. (2004). Outdoor adventure in promoting relationships with nature. *Journal of Outdoor and Environmental Education*, 8(1), 20–8.

17. Pirchio, S., Passiatore, Y., Panno, A., et al. (2021). The effects of contact with nature during outdoor environmental education on students' wellbeing, connectedness to nature and pro-sociality. *Frontiers in Psychology*, 1523.

18. Ernst, J., & Theimer, S. (2011). Evaluating the effects of environmental education programming on connectedness to nature. *Environmental Education Research*, 17(5), 577–98; Liefländer, A. K., Fröhlich, G., Bogner, F. X., & Schultz, P. W. (2013). Promoting connectedness with nature through environmental education. *Environmental Education Research*, 19(3), 370–84.

19. Lankenau, G. R. (2018). Fostering connectedness to nature in higher education. *Environmental Education Research*, 24(2), 230–44.

20. Nisbet, E. K., Zelenski, J. M., & Murphy, S. A. (2011). Happiness is in our nature: exploring nature relatedness as a contributor to subjective well-being. *Journal of Happiness Studies*, 12(2), 303–22.

21. Bruni, C. M., Winter, P. L., Schultz, P. W., et al. (2017). Getting to know nature: evaluating the effects of the Get to Know Program on children's connectedness with nature. *Environmental Education Research*, 23(1), 43–62.

22. Sellmann, D., & Bogner, F. X. (2013). Effects of a 1-day environmental education intervention on environmental attitudes and connectedness with nature. *European Journal of Psychology of Education*, 28(3), 1077–86.

23. Braun, T., & Dierkes, P. (2017). Connecting students to nature – how intensity of nature experience and student age influence the success of outdoor education programs. *Environmental Education Research*, 23(7), 937–49.

24. Kossack, A., & Bogner, F. X. (2012). How does a one-day environmental education programme support individual connectedness with nature? *Journal of Biological Education*, 46(3), 180–7.

25. Barragan-Jason, G., de Mazancourt, C., Parmesan, C., et al. (2021). Human–nature connectedness as a pathway to sustainability: a global meta-analysis. *Conservation Letters*, e12852.

26. Eisenstein, C. (2013). *The Ascent of Humanity: Civilization and the Human Sense of Self*. Berkeley, CA: North Atlantic Books.

27. Barrable, A., & Booth, D. (2020). Increasing nature connection in children: a mini review of interventions. *Frontiers in Psychology*, 11, 492.

28. Buijs, A., & Lawrence, A. (2013). Emotional conflicts in rational forestry: towards a research agenda for understanding emotions in environmental conflicts. *Forest Policy and Economics*, 33, 104–11.

29. Kellert, S. R., & Wilson, E. O. (1993). *The Biophilia Hypothesis. Island Press, Washington, DC*.

30. Lumber, R., Richardson, M., & Sheffield, D. (2017). Beyond knowing nature: contact, emotion, compassion, meaning, and beauty are pathways to nature connection. *PLoS One*, 12(5), e0177186. https://doi.org/10.1371/journal.pone.0177186

31. Lumber, R., Richardson, M., & Sheffield, D. (2017). Beyond knowing nature.

32. Lumber, R., Richardson, M., & Sheffield, D. (2017). Beyond knowing nature; Richardson, M., Dobson, J., Abson, D. J., et al. (2020). Applying

the pathways to nature connectedness at a societal scale: a leverage points perspective. *Ecosystems and People*, 16(1), 387–401; Richardson, M. (2020). Applying the Pathways to Nature Connectedness at Societal Scale. Applying the Pathways to Nature Connectedness at Societal Scale | Finding Nature

33. Capaldi, C. A., Passmore, H. A., Ishii, R., et al. (2017). Engaging with natural beauty may be related to well-being because it connects people to nature: evidence from three cultures. *Ecopsychology*, 9(4), 199–211.

34. Charlton, N. G. (2008). *Understanding Gregory Bateson: Mind, Beauty, and the Sacred Earth*. Albany, NY: SUNY Press.

35. Shonin, E., Van Gordon, W., Singh, N. N., & Griffiths, M. D. (2015). Mindfulness of emptiness and the emptiness of mindfulness. In E. Shonin, W. Van Gordon, N. N. Singh (eds) *Buddhist Foundations of Mindfulness* (pp. 159–78). Cham: Springer.

36. Richardson, M., Cormack, A., McRobert, L., & Underhill, R. (2016). 30 Days Wild: development and evaluation of a large-scale nature engagement campaign to improve well-being. *PloS One*, 11(2), e0149777; Richardson, M., & McEwan, K. (2018). 30 Days Wild and the relationships between engagement with nature's beauty, nature connectedness and well-being. *Frontiers in Psychology*, 9, 1500; Richardson, M., McEwan, K., & Garip, G. (2018). 30 Days Wild: who benefits most? *Journal of Public Mental Health*; 30 Days Wild 5 Year Review | The Wildlife Trusts. (n.d.). www.wildlifetrusts.org. https://www.wildlifetrusts.org/30-days-wild-5-year-review

37. White, M., Hamlin, I., Butler, C.W. & Richardson, M. (Under review). The Joy of Birds: The effect of rating for joy or counting garden bird species on wellbeing, anxiety, and nature connection. *Urban Ecosystems*.

38. Shutt, J. D., & Lees, A. C. (2021). Killing with kindness: does widespread generalised provisioning of wildlife help or hinder biodiversity conservation efforts? *Biological Conservation*, 261, 109295.

39. Muneghina, O., Van Gordon, W., Barrows, P., & Richardson, M. (2021). A novel mindful nature connectedness intervention improves paranoia but not anxiety in a nonclinical population. *Ecopsychology*. https://doi.org/10.1089/eco.2020.0068

40. Van der Gaag, M., Valmaggia, L. R., & Smit, F. (2014). The effects of individually tailored formulation-based cognitive behavioural therapy in auditory hallucinations and delusions: a meta-analysis. *Schizophrenia Research*, 156(1), 30–7.

41. Rhodes, C., & Lumber, R. (2021). Using the five pathways to ature to make a spiritual connection in early recovery from SUD: a pilot study. *International Journal of Mental Health and Addiction*. https://doi.org/10.1007/s11469-021-00565-4

42. Johnson, J. E., Finney, J. W., & Moos, R. H. (2006). End-of-treatment outcomes in cognitive-behavioral treatment and 12-step substance use treatment programs: do they differ and do they predict 1-year outcomes? *Journal of Substance Abuse Treatment*, 31(1), 41–50.

43.   Schoenthaler, S. J., Blum, K., Braverman, E. R., et al. (2015). NIDA-Drug Addiction Treatment Outcome Study (DATOS) relapse as a function of spirituality/religiosity. *Journal of Reward Deficiency Syndrome*, 1(1), 36; Dermatis, H., & Galanter, M. (2016). The role of twelve-step-related spirituality in addiction recovery. *Journal of Religion and Health*, 55(2), 510–21.

44.   Tonigan, J. S., Rynes, K. N., & McCrady, B. S. (2013). Spirituality as a change mechanism in 12-step programs: a replication, extension, and refinement. *Substance Use & Misuse*, 48(12), 1161–73.

45.   Kamitsis, I., & Francis, A. J. (2013). Spirituality mediates the relationship between engagement with nature and psychological wellbeing. *Journal of Environmental Psychology*, 36, 136–43.

46.   Fanning, A. L., O'Neill, D. W., Hickel, J., & Roux, N. (2021). The social shortfall and ecological overshoot of nations. *Nature Sustainability*, 1–11.

47.   Twenge, J. M., Miller, J. D., & Campbell, W. K. (2014). The narcissism epidemic: commentary on modernity and narcissistic personality disorder. *Personality Disorders: Theory, Research, and Treatment*, 5(2), 227–9; Logan, A. C., & Prescott, S. L. (2022). Planetary health: we need to talk about narcissism. *Challenges*, 13(1), 19; Van Gordon, W., Shonin, E., Diouri, S., et al. (2018). Ontological addiction theory: attachment to me, mine, and I. *Journal of Behavioral Addictions*, 7(4), 892–6.

Chapter 12: Scaling Up: Policies for Connection

1.   Wells, V. K. (2012). Foraging: an ecology model of consumer behaviour? *Marketing Theory*, 12(2), 117–36; Wells, V. K. (2014). Behavioural psychology, marketing and consumer behaviour: a literature review and future research agenda. *Journal of Marketing Management*, 30(11–12), 1119–58.

2.   Aminpour, P., Gray, S. A., Beck, M. W., et al. (2022). Urbanized knowledge syndrome – erosion of diversity and systems thinking in urbanites' mental models. *npj Urban Sustainability*, 2(1), 1–10.

3.   Davis, M. C., Challenger, R., Jayewardene, D. N., & Clegg, C. W. (2014). Advancing socio-technical systems thinking: a call for bravery. *Applied Ergonomics*, 45(2), 171–80.

4.   Catchpole, K., Bowie, P., Fouquet, S., et al. (2021). Frontiers in human factors: embedding specialists in multi-disciplinary efforts to improve healthcare. *International Journal for Quality in Health Care*, 33(Supplement_1), 13–18.

5.   Evbuoma, E. I., Hu, M., Farrell, A., et al. (2021). *Systems Thinking Iceberg: Diving Beneath the Surface in Education Systems.* Social System Design Lab, Washington University in St. Louis.

6.   Lorenz, E. (2000). The butterfly effect. *World Scientific Series on Nonlinear Science Series A*, 39, 91–4.

7.   Ingold, T. (2000). *The Perception of the Environment: Essays on Livelihood, Dwelling and Skill.* London: Routledge.

8.   Aminpour, P., Gray, S. A., Beck, M. W., et al. (2022). Urbanized knowledge syndrome.

9.   Engel, G. L. (1977). The need for a new medical model: a challenge for biomedicine. *Science*, 196(4286), 129–36.

10. The Unfrozen Moment – Delivering A Green Brexit. (n.d.). GOV.UK. https://www.gov.uk/government/speeches/the-unfrozen-moment-delivering-a-green-brexit

11. Gross, J. J. (2013). Emotion regulation: taking stock and moving forward. *Emotion*, 13(3), 359; Skov, M., & Nadal, M. (2020). The nature of beauty: behavior, cognition, and neurobiology. https://doi.org/10.31234/osf.io/c5m87

12. Eisenstein, C. (2013). *The Ascent of Humanity: Civilization and the Human Sense of Self*. Berkeley, CA: North Atlantic Books.

13. Landscapes review: National Parks and AONBs. (n.d.). GOV.UK. https://www.gov.uk/government/publications/designated-landscapes-national-parks-and-aonbs-2018-review

14. Richardson, M., Richardson, E., Hallam, J., & Ferguson, F. J. (2020). Opening doors to nature: bringing calm and raising aspirations of vulnerable young people through nature-based intervention. *The Humanistic Psychologist*, 48(3), 284.

15. DEFRA. (2018). 25 Year Environment Plan. GOV.UK. https://www.gov.uk/government/publications/25-year-environment-plan

16. Eastwood, A., Juárez-Bourke, A., Herrett, S., & Hague, A. (2021). Connecting young people with greenspaces: the case for participatory video. *People and Nature*. https://doi.org/10.1002/pan3.10236

17. PIRC. (2021). Framing Nature Toolkit. PIRC. https://publicinterest.org.uk/nature-toolkit/

18. Green Recovery Challenge Fund round 1 | The National Lottery Heritage Fund. (2018). www.heritagefund.org.uk. https://www.heritagefund.org.uk/funding/closed-programmes/green-recovery-challenge-fund#assess

19. Connect with nature | Generation Green. (n.d.). www.yha.org.uk. https://www.yha.org.uk/generationgreen

20. Building partnerships for Nature's recovery. (n.d.). GOV.UK. https://www.gov.uk/government/publications/natural-england-building-partnerships-for-natures-recovery/building-partnerships-for-natures-recovery

21. Public Health England. (2020). Improving access to greenspace A new review for 2020. https://assets.publishing.service.gov.uk/government/uploads/system/uploads/attachment_data/file/904439/Improving_access_to_greenspace_2020_review.pdf

22. Government unveils levelling up plan that will transform UK. (2022). GOV.UK. https://www.gov.uk/government/news/government-unveils-levelling-up-plan-that-will-transform-uk

23. Nature and the City. (n.d.). Policy Exchange. https://policyexchange.org.uk/publication/nature-and-the-city/

24. António Guterres, Secretary-General of the United Nations, in United Nations Environment Programme (2021). *Making Peace with Nature: A Scientific Blueprint to Tackle the Climate, Biodiversity and Pollution Emergencies.*

25. SEI & CEEW (2022). *Stockholm+50: Unlocking a Better Future.* Stockholm Environment Institute. https://doi.org/10.51414/sei2022.011

26. Capaldi, C. A., Dopko, R. L., & Zelenski, J. M. (2014). The relationship between nature connectedness and happiness: a meta-analysis. *Frontiers in Psychology*, 5, 976; Dasgupta, P. (2021), The Economics of Biodiversity: The Dasgupta Review. (London: HM Treasury).

27. Dasgupta, P. (2021), The Economics of Biodiversity.

28. Nature, biodiversity and health: an overview of interconnections. Copenhagen: WHO Regional Ofce for Europe; 2021. Licence: CC BY-NC-SA 3.0 IGO. https://www.euro.who.int/en/publications/abstracts/nature,-biodiversity-and-health-an-overview-of-interconnections-2021

29. Næss, A. (1973). The shallow and the deep, long-range ecology movement. A summary. *Inquiry*, 16(1–4), 95-100.

30. A Deeper Appreciation of Nature. (n.d.). www.resurgence.org. https://www.resurgence.org/magazine/article4982-a-deeper-appreciation-of-nature.html

31. West, S., Haider, L. J., Stålhammar, S., & Woroniecki, S. (2020). A relational turn for sustainability science? Relational thinking, leverage points and transformations. *Ecosystems and People*, 16(1), 304–25.

32. Mental Health Foundation. (2021). *Nature. How connecting with nature benefits our mental health.* https://www.mentalhealth.org.uk/sites/default/files/2022-06/MHAW21-Nature-research-report.pdf

33. Mental Health Foundation. (2021). *Mental Health and Nature Policy briefing.* https://www.mentalhealth.org.uk/sites/default/files/2022-06/MHF-Mental-Health-and-Nature-Policy-Briefing-MHAW-2021.pdf

34. Richardson, M. (2021). How can "nature connectedness" improve wellbeing for people and nature? What Works Wellbeing. https://whatworkswellbeing.org/blog/how-can-nature-connectedness-improve-wellbeing-for-people-and-nature/; Richardson, M. (2021). Moments, not Minutes: The nature-wellbeing link. Behavioural Public Policy Blog. https://bppblog.com/2021/07/09/moments-not-minutes-the-nature-wellbeing-link/?amp=1

Chapter 13: Tools for Change

1. António Guterres, Secretary-General of the United Nations, in United Nations Environment Programme (2021). *Making Peace with Nature: A Scientific Blueprint to Tackle the Climate, Biodiversity and Pollution Emergencies.*

2. IPBES Secretariat. (2019). Transformative change assessment. IPBES Secretariat. https://ipbes.net/transformative-change; ISDD. (n.d.).

IPBES to Assess Transformational Changes to Tackle Biodiversity Loss | News | SDG Knowledge Hub | IISD. https://sdg.iisd.org/news/ipbes-to-assess-transformational-changes-to-tackle-biodiversity-loss/

3.   Hirvilammi, T., & Helne, T. (2014). Changing paradigms: a sketch for sustainable wellbeing and ecosocial policy. *Sustainability*, 6(4), 2160–75.

4.   Meadows, D. H. (1999). *Leverage Points: Places to Intervene in a System.* https://donellameadows.org/archives/leverage-points-places-to-intervene-in-a-system

5.   Star, S. L., & Griesemer, J. R. (1989). Institutional ecology, 'translations' and boundary objects: amateurs and professionals in Berkeley's Museum of Vertebrate Zoology, 1907–39. *Social Studies of Science*, 19(3), 387–420.

6.   Abson, D. J., Fischer, J., Leventon, J., et al. (2017). Leverage points for sustainability transformation. *Ambio*, 46(1), 30–9.

7.   Meadows, D. H. (1999). *Leverage Points: Places to Intervene in a System.*

8.   Fischer, J., & Riechers, M. (2019). A leverage points perspective on sustainability. *People and Nature*, 1(1), 115–20.

9.   Douglas, J. W., & Evans, K. L. (2021). An experimental test of the impact of avian diversity on attentional benefits and enjoyment of people experiencing urban green-space. *People and Nature*. https://doi.org/10.1002/pan3.10279; Batool, A., Rutherford, P., McGraw, P., et al. (2021). Gaze correlates of view preference: comparing natural and urban scenes. *Lighting Research & Technology*, 14771535211055703; Bingjing, C., Chen, G., & Shuhua, L. I. (2022). Looking at buildings or trees? Association of human nature relatedness with eye movements in outdoor space. *Journal of Environmental Psychology*, 101756.

10. Goldsmith, B. (2022) How to restore the British countryside. *Spectator.* https://www.spectator.co.uk/article/how-to-restore-the-british-countryside

11. Buijs, A., Kamphorst, D., Mattijssen, T., et al. (2022). Policy discourses for reconnecting nature with society: the search for societal engagement in Dutch nature conservation policies. *Land Use Policy*, 114, 105965.

12. Goldsmith, B. (2022) How to restore the British countryside.

13. Meadows, D. H. (1999). *Leverage Points.*

14. Fischer, J., & Riechers, M. (2019). A leverage points perspective on sustainability.

15. Kuhn, T. S. (1962). *The Structure of Scientific Revolutions.* Chicago: University of Chicago Press.

16. Tree, I. (2018). *Wilding: The Return of Nature to a British Farm.* London: Pan Macmillan.

17. Allestree Park: Calls for mountain bike trails as part of new plans. (2022). BBC News. https://www.bbc.co.uk/news/uk-england-derbyshire-60379882

18. Tree, I. Personal communication, 8 March 2022.

19. Abson, D. J., Fischer, J., Leventon, J., et al. (2017). Leverage points for sustainability transformation. *Ambio*, 46(1), 30–9; Dorninger, C., Abson, D. J., Apetrei, C. I., et al. (2020). Leverage points for sustainability

transformation: a review on interventions in food and energy systems. *Ecological Economics*, 171, 106570; Ehrenfeld, J. R. (2004). Searching for sustainability: no quick fix. *Reflections*, 5(8), 1–13.

20. Eisenstein, C. (2013). *The Ascent of Humanity: Civilization and the Human Sense of Self*. Berkeley, CA: North Atlantic Books

21. Riechers, M., Balázsi, Á., García-Llorente, M., & Loos, J. (2021). Human–nature connectedness as leverage point. *Ecosystems and People*, 17(1), 215–21; Riechers, M., Loos, J., Balázsi, Á., et al. (2021). Key advantages of the leverage points perspective to shape human–nature relations. *Ecosystems and People*, 17(1), 205–14. https://doi.org/10.1080/2639591 6.2021.1912829

22. Abson, D. J., Fischer, J., Leventon, J., et al. (2017). Leverage points for sustainability transformation.

23. Richardson, M., Dobson, J., Abson, D. J., et al. (2020). Applying the pathways to nature connectedness at a societal scale: a leverage points perspective. *Ecosystems and People*, 16(1), 387–401.

24. Lakin, J. L., Jefferis, V. E., Cheng, C. M., & Chartrand, T. L. (2003). The chameleon effect as social glue: evidence for the evolutionary significance of nonconscious mimicry. *Journal of Nonverbal Behavior*, 27(3), 145–62; Cacioppo, J. T., & Patrick, W. (2008). *Loneliness: Human Nature and the Need for Social Connection*. New York: WW Norton & Company; Gilbert, P. (2014). The origins and nature of compassion focused therapy. *British Journal of Clinical Psychology*, 53(1), 6–41.

25. Tam, K. P., Lee, S. L., & Chao, M. M. (2013). Saving Mr. Nature: anthropomorphism enhances connectedness to and protectiveness toward nature. *Journal of Experimental Social Psychology*, 49(3), 514–21.

26. Ives, C. D., Abson, D. J., von Wehrden, H., et al. (2018). Reconnecting with nature for sustainability. *Sustainability Science*, 13(5), 1389–97.

27. Haidt, J., Keltner, D., Peterson, C., & Seligman, M. E. (2004). Appreciation of beauty and excellence. *Character Strengths and Virtues*, 537–51.

28. Diessner, R., Solom, R. D., Frost, N. K., et al. (2008). Engagement with beauty: appreciating natural, artistic, and moral beauty. *Journal of Psychology*, 142(3), 303–32.

29. Fleming, M. (2018). Why Cadbury's 'Gorilla' ad nearly didn't get made. *Marketing Week*. https://www.marketingweek.com/cadbury-gorilla/

30. Holbrook, M. B., & Hirschman, E. C. (1982). The experiential aspects of consumption: consumer fantasies, feelings, and fun. *Journal of Consumer Research*, 9(2), 132–40; Stieler, M., & Germelmann, C. C. (2016). The ties that bind us: feelings of social connectedness in socio-emotional experiences. *Journal of Consumer Marketing*.

31. Pine, B. J., Pine, J., & Gilmore, J. H. (1999). *The Experience Economy: Work is Theatre & Every Business a Stage*. Boston: Harvard Business School Press.

32. Baumeister, R. F. (1991). *Meanings of Life*. New York: Guilford.

33. Park, C. L. (2010). Making sense of the meaning literature: an integrative review of meaning making and its effects on adjustment to stressful life events. *Psychological Bulletin*, 136(2), 257.

34.   Kenter, J. O., O'Brien, L., Hockley, N., et al. (2015). What are shared and social values of ecosystems? *Ecological Economics*, 111, 86–99.

35.   Arnould, E. J., & Price, L. L. (1993). River magic: extraordinary experience and the extended service encounter. *Journal of Consumer Research*, 20(1), 24–45; Holt, D. B. (1995). How consumers consume: a typology of consumption practices. *Journal of Consumer Research*, 22(1), 1–16; Keltner, D., & Haidt, J. (1999). Social functions of emotions at four levels of analysis. *Cognition & Emotion*, 13(5), 505–21; Parkinson, B. (1996). Emotions are social. *British Journal of Psychology*, 87(4), 663–83.

36.   Karpinska-Krakowiak, M., & Eisend, M. (2021). The effects of animistic thinking, animistic cues, and superstitions on brand responses on social media. *Journal of Interactive Marketing*, 55, 104–17; Chandler, J., & Schwarz, N. (2010). Use does not wear ragged the fabric of friendship: thinking of objects as alive makes people less willing to replace them. *Journal of Consumer Psychology*, 20(2), 138–45.

37.   Holbrook, M. B., & Hirschman, E. C. (1982). The experiential aspects of consumption.

38.   Wolf, K. L. (2005). Business district streetscapes, trees, and consumer response. *Journal of Forestry*, 103(8), 396–400.

39.   Kurzman, C. (2008). Meaning-making in social movements. *Anthropological Quarterly*, 81(1), 5–15.

40.   Klein, E. (2020). How to topple dictators and transform society. *Vox*. https://www.vox.com/podcasts/2020/1/3/21048121/ezra-klein-erica-chenoweth-nonviolence-topple-dictators

41.   Nardini, G., Rank-Christman, T., Bublitz, M. G., et al. (2021). Together we rise: how social movements succeed. *Journal of Consumer Psychology*, 31(1), 112–45; Chenoweth, E., Stephan, M. J., & Stephan, M. J. (2011). *Why Civil Resistance Works: The Strategic Logic of Nonviolent Conflict*. New York: Columbia University Press.

42.   Beckerman, G. (2022). *The Quiet Before*. London: Bantam.

43.   Collins, R. (2001). Social movements and the focus of emotional attention. In J. Goodwin, J. M. Jasper & F. Polletta (eds) *Passionate Politics: Emotions and Social Movements* (pp. 27–44). Chicago: University of Chicago Press; Kurzman, C. (2008). Meaning-making in social movements. *Anthropological Quarterly*, 81(1), 5–15.

44.   Moyer, B. (1987). *The Movement Action Plan: A Strategic Framework Describing the Eight Stages of Successful Social Movements*; https://commonslibrary.org/resource-bill-moyers-movement-action-plan/

45.   Extinction Rebellion. (2022). About Us. Extinction Rebellion UK. https://extinctionrebellion.uk/the-truth/about-us/

46.   Nardini, G., Rank-Christman, T., Bublitz, M. G., et al. (2021). Together we rise: how social movements succeed. *Journal of Consumer Psychology*, 31(1), 112–45.

Chapter 14: Creating a Nature-Connected Society

1.   Angle, S. C. (2009). *Sagehood: The Contemporary Significance of Neo-Confucian Philosophy*. New York: Oxford University Press.

2.   Eisenstein, C. (2013). *The Ascent of Humanity: Civilization and the Human Sense of Self*. Berkeley, CA: North Atlantic Books.

3.   Carl, J. (2009). Industrialization and public education: social cohesion and social stratification. In R. Cowen & A. M. Kazamias (eds) *International Handbook of Comparative Education* (pp. 503–18). Dordrecht: Springer.

4.   GOV.UK. (2022). About us. GOV.UK. https://www.gov.uk/government/organisations/department-for-education/about

5.   DfE (2022). *Sustainability and Climate Change: A Strategy for the Education & Children's Services Systems*. https://www.gov.uk/government/publications/sustainability-and-climate-change-strategy/sustainability-and-climate-change-a-strategy-for-the-education-and-childrens-services-systems; Dunlop, L. & Rushton, E. A. C. (2022). Putting climate change at the heart of education: is England's strategy a placebo for policy? *British Educational Research Journal*, 1–19. https://doi.org/10.1002/berj.3816

6.   Abson, D. J., Fischer, J., Leventon, J., et al. (2017). Leverage points for sustainability transformation. *Ambio*, 46(1), 30–9.

7.   Hongxin, W., Khan, M. A., Zhenqiang, J., et al. (2022). Unleashing the role of CSR and employees' pro-environmental behavior for organizational success: the role of connectedness to nature. *Sustainability*, 14(6), 3191.

8.   Karami, A., & Gorzynski, R. A. (2021). Connection to nature and sustainability in small- and medium-sized environmental organizations: a dynamic strategic thinking approach. *Business Strategy and the Environment*. https://doi.org/10.1002/bse.2898

9.   Leading for Wellbeing. (n.d.). *Meadows Memorandum: A New Economic Model for a Finer Future*. https://wellbeingeconomy.org/wp-content/uploads/2019/05/Meadows-Memorandum-with-Cover-V81-copy.pdf

10. Petrescu, D., Petcou, C., Safri, M., & Gibson, K. (2021). Calculating the value of the commons: generating resilient urban futures. *Environmental Policy and Governance*, 31(3), 159–74.

11. SEI & CEEW (2022). *Stockholm+50: Unlocking a Better Future*. Stockholm Environment Institute. https://doi.org/10.51414/sei2022.011; Richardson, M., Dobson, J., Abson, D. J., et al. (2020). Applying the pathways to nature connectedness at a societal scale: a leverage points perspective. *Ecosystems and People*, 16(1), 387–401, https://doi.org/10.1080/26395916.2020.1844296

12. National Trust encourages people to celebrate spring with UK's first ever #BlossomWatch day. (2022). National Trust. https://www.nationaltrust.org.uk/press-release/national-trust-encourages-people-to-celebrate-spring-with-uks-first-ever-blossomwatch-day

13. Shrubsole, G. (2019). *Who Owns England? How We Lost Our Green and Pleasant Land, and How to Take It Back*. London: HarperCollins.

14. Cocker, M. (2018). *Our Place: Can we Save Britain's Wildlife Before It Is Too Late?* London: Random House.

15. Restall, B., Conrad, E. & Cop, C. (2021). Connectedness to nature: mapping the role of protected areas. *Journal of Environmental Management*, 293, 112771.

16. García-Llorente, M., Pérez-Ramírez, I., Sabán de la Portilla, C., et al. (2019). Agroecological strategies for reactivating the agrarian sector: the case of Agrolab in Madrid. *Sustainability*, 11(4), 1181; Pérez-Ramírez, I., García-Llorente, M., Saban de la Portilla, C., et al. (2021). Participatory collective farming as a leverage point for fostering human–nature connectedness. *Ecosystems and People*, 17(1), 222–34.

17. Leiserowitz, A. A., Maibach, E. W., Roser-Renouf, C., et al. (2013). Climategate, public opinion, and the loss of trust. *American Behavioral Scientist*, 57(6), 818–37; McKie, R. (2019, November 9). Climategate 10 years on: what lessons have we learned? The Guardian. https://www.theguardian.com/theobserver/2019/nov/09/climategate-10-years-on-what-lessons-have-we-learned

18. Rose, L. (2020). *Framing Nature: Conservation and Culture*. Hebden Bridge: Gritstone Publishing Co-operative.

19. Rewilding People & Places. (n.d.). Greenminds. https://greenmindsplymouth.com/

20. Down to Earth Derby - Nature-Based Regeneration. (n.d.). www.dtederby.org. https://www.dtederby.org/

21. Butler, T. (2020). Making the Museum of Making at Derby Silk Mill. In G. Black (ed.), *Museums and the Challenge of Change: Old Institutions in a New World*. Abingdon: Routledge.

22. Neaverson, P., & Palmer, M. (2002). *Industry in the Landscape, 1700–1900*. Abingdon: Routledge.

23. Albritton Jonsson, F. (2012). The Industrial Revolution in the Anthropocene. *Journal of Modern History*, 84(3), 679–96.

Chapter 15: Designing a Connected Future

1. Richardson, M. (2008). Ergonomists as choice architects: behavioural economics and ergonomics. *The Ergonomist*, 458, 1–2.

2. Maier, M., Bartoš, F., Stanley, T. D., Shanks, D. R., Harris, A. J., & Wagenmakers, E. J. (2022). No evidence for nudging after adjusting for publication bias. *Proceedings of the National Academy of Sciences*, 119(31), e2200300119.

3. Thaler, R. H., & Sunstein, C. R. (2008). *Nudge: Improving Decisions about Health, Wealth, and Happiness*. New Haven, CT: Yale University Press.

4. Gibson, J. J. (1979). *The Ecological Approach to Visual Perception*. New York: Houghton Mifflin.

5. Laaksoharju, T., & Rappe, E. (2017). Trees as affordances for connectedness to place – a framework to facilitate children's relationship with nature. *Urban Forestry & Urban Greening*, 28, 150–9.

6. Kaaronen, R. O., & Strelkovskii, N. (2020). Cultural evolution of sustainable behaviors: pro-environmental tipping points in an agent-based model. *One Earth*, 2(1), 85–97.

7. Linder, N., Giusti, M., Samuelsson, K., & Barthel, S. (2021). Pro-environmental habits: an underexplored research agenda in sustainability science. *Ambio*, 1–11.

8. Raymond, C. M., Kyttä, M., & Stedman, R. (2017). Sense of place, fast and slow: the potential contributions of affordance theory to sense of place. *Frontiers in Psychology*, 8, 1674.

9. Kellert, S., & Calabrese, E. (2015). *The Practice of Biophilic Design*. London: Terrapin Bright LLC.

10. Hung, S. H., & Chang, C. Y. (2021). Health benefits of evidence-based biophilic-designed environments: a review. *Journal of People, Plants, and Environment*, 24(1), 1–16; Gillis, K., & Gatersleben, B. (2015). A review of psychological literature on the health and wellbeing benefits of biophilic design. *Buildings*, 5(3), 948–63.

11. Richardson, M., & Butler, C. W. (2022). Nature connectedness and biophilic design. *Building Research & Information*, 50(1–2), 36–42.

12. Hari, J. (2022). Your attention didn't collapse. It was stolen. The Guardian. https://www.theguardian.com/science/2022/jan/02/attention-span-focus-screens-apps-smartphones-social-media; Hari, J. (2022). *Stolen Focus: Why You Can't Pay Attention*. London: Bloomsbury; Douglas, J. W., & Evans, K. L. (2021). An experimental test of the impact of avian diversity on attentional benefits and enjoyment of people experiencing urban green-space. *People and Nature*. https://doi.org/10.1002/pan3.10279

13. Chalmers, D. J. (2022). *Reality+: Virtual Worlds and the Problems of Philosophy*. London: Allen Lane.

14. Taylor, T. (2010). *The Artificial Ape: How Technology Changed the Course of Human Evolution*. New York: St. Martin's Press.

15. Gray, J. (2002). *Straw Dogs: Thoughts on Humans and Other Animals*. New York: Farrar, Straus and Giroux.

16. Gray, J. (2002). *Straw Dogs*.

17. Campbell, C. (2022). *AI by Design: A Plan for Living with Artificial Intelligence*. Boca Raton, FL: Chapman and Hall/CRC.

18. Breazeal, C. L. (2002). *Designing Sociable Robots*. Cambridge, MA: MIT Press.

19. Broekens, J., Heerink, M., & Rosendal, H. (2009). Assistive social robots in elderly care: a review. *Gerontechnology*, 8(2), 94–103; McGlynn, S. A., Kemple, S., Mitzner, T. L., et al. (2017). Understanding the potential of PARO for healthy older adults. *International Journal of Human–Computer Studies*, 100, 33–47; Jøranson, N., Pedersen, I., Rokstad, A. M. M., & Ihlebaek, C. (2015). Effects on symptoms of agitation and depression in persons with dementia participating in robot-assisted activity: a cluster-randomized controlled trial. *Journal of the American Medical Directors Association*, 16(10), 867–73; Hudson, J., Ungar, R., Albright, L., et al. (2020). Robotic pet use among community-dwelling older adults. *Journals of Gerontology: Series B*, 75(9), 2018–28.

20. Burrows, A. M., & Omstead, K. M. (2022). Dog faces Are faster than wolf faces. *FASEB Journal*, 36. https://doi.org/10.1096/fasebj.2022.36. S1.R5001

21. Harari, Y. N. (2016). *Homo Deus: A Brief History of Tomorrow*. New York: Random House.

22. Nightingale, S. J., & Farid, H. (2022). AI-synthesized faces are indistinguishable from real faces and more trustworthy. *Proceedings of the National Academy of Sciences*, 119(8). https://doi.org/10.1073/pnas.2120481119

23. Gray, J. (2002). *Straw Dogs*.

24. An Ecomodernist Manifesto. (n.d.). The Breakthrough Institute. https://thebreakthrough.org/articles/an-ecomodernist-manifesto

25. Zimmer, C. (2021). A New Company With a Wild Mission: Bring Back the Woolly Mammoth. *New York Times*. https://www.nytimes.com/2021/09/13/science/colossal-woolly-mammoth-DNA.html

26. Hamilton, C. (2016). The theodicy of the 'Good Anthropocene'. *Environmental Humanities*, 7(1), 233–8.

27. Lovelock, J. (2019). Novacene: The *Coming Age of Hyperintelligence*. Cambridge, MA: MIT Press.

28. Wong, M. L., & Bartlett, S. (2022). Asymptotic burnout and homeostatic awakening: a possible solution to the Fermi paradox? *Journal of the Royal Society Interface*, 19(190), 20220029.

29. Gigerenzer, G. (2022). *How to Stay Smart in a Smart World*. London: Allen Lane.

30. Opportunities in the metaverse. (n.d.). https://www.jpmorgan.com/content/dam/jpm/treasury-services/documents/opportunities-in-the-metaverse.pdf; The metaverse will steal your identity. (2022, March 3). UnHerd. https://unherd.com/2022/03/the-metaverse-will-steal-your-identity/

31. Gray, J. (2002). *Straw Dogs*.

32. Gray, J. (2002). *Straw Dogs*.

33. Kings College London. (2022). Are attention spans really collapsing? Data shows UK public are worried – but also see benefits from technology. https://www.kcl.ac.uk/news/are-attention-spans-really-collapsing-data-shows-uk-public-are-worried-but-also-see-benefits-from-technology

Index

Miles Richardson is Professor of Human Factors and Nature Connectedness at the University of Derby, where he founded the Nature Connectedness Research Group. He has pioneered widely adopted and award-winning approaches to improving the human–nature relationship. Author of dozens of scientific papers, he advises nationally and internationally on uniting human and nature's wellbeing.

Blog: findingnature.org.uk

Twitter: @findingnature

Instagram: findingnature_